Structural Health Monitoring for Suspension Bridges

Yang Deng · Aiqun Li

Structural Health Monitoring for Suspension Bridges

Interpretation of Field Measurements

Science Press
Beijing

Springer

Yang Deng
Beijing Advanced Innovation
Center for Future Urban Design
Beijing University of Civil Engineering
and Architecture
Beijing, China

Aiqun Li
Beijing Advanced Innovation
Center for Future Urban Design
Beijing University of Civil Engineering
and Architecture
Beijing, China

ISBN 978-981-13-3346-0 ISBN 978-981-13-3347-7 (eBook)
https://doi.org/10.1007/978-981-13-3347-7

Jointly published with Science Press, Beijing, China

The print edition is not for sale in China Mainland. Customers from China Mainland please order the print book from: Science Press, Beijing, China.

Library of Congress Control Number: 2018962387

This Springer imprint is published by the registered company Springer Nature Singapore Pte Ltd.
The registered company address is: 152 Beach Road, #21-01/04 Gateway East, Singapore 189721, Singapore

Preface

The Motivation for This Book

The past decades have witnessed the emergence of applying structural health monitoring (SHM) system to monitor and evaluate the healthy condition of civil infrastructures. China also has made great progresses in bridge SHM with the rapid and large-scale construction of transportation networks. During the past two decades, hundreds of SHM systems have been established for long-span bridges, especially suspension bridges.

The main motivation of this book stems from the observation that although many sensing techniques and data analysis methods have been deployed, some real requirements in SHM of suspension bridges have not been fully addressed. Most of newly built suspension bridges in China have been instrumented with hundreds of sensors to collect information of loads and structural responses. However, the massive data produced by the costly SHM systems usually does not receive enough analysis. Second, the structural behavior of the suspension bridges is pretty complicated due to complex configuration and in-service loading conditions. The comprehensive and reasonable understanding of structural behaviors and performance of the suspension bridges call for deep and enough interpretation on multisource measurements by using both modern signal process techniques and conventional structural theories.

In this book, we present our 10-year research achievements in field measurement interpretation for the suspension bridges. Some frameworks are especially designed to address the unique requirements of SHM applications for suspension bridges. We share our experiences and findings by stepping from sensing technique selection and data processing to performance assessment for the long-span suspension bridges.

What This Book is About

This book provides comprehensive coverage and detailed insights into the interpretation of field measurements of using SHM for the suspension bridges. The readers can benefit from this book in sensing technique selection, data processing, and performance assessment. In particular, this book helps the readers to develop the frameworks of field measurement interpretation when they face massive and multisource monitoring information from the SHM of long-span bridges. Therefore, it can be seen as a practical guide as well as a textbook for the reader:

- To establish the measurement interpretation frameworks in domain-specific SHM applications for the suspension bridges.
- To learn about the new insights of structural performance and behaviors of the suspension bridges from the authors' practice and experience of SHM for the long-span suspension bridges.

How This Book is Organized

This book is divided into five parts.

The first part includes Chaps. 1 "Temperature Action Monitoring of Main Girder" and 2 "Bridge-Site Extreme Wind Prediction". This part presents the monitoring results of two important environmental actions for long-span suspension bridges, including temperature action and bridge-site wind. Bridge designers can learn from this part to improve the potential bridge design.

The second part consists of Chaps. 3 "Measurement-Based Damage Detection for Expansion Joints" and 4 "Modal Frequency-Based Structural Damage Detection". Damage detection is one of the main topics in SHM. However, the structural responses or inherent features of the suspension bridges are not sensitive to the local structural damages. The changes of damage detection indices caused by structural damages are easily masked by the environmental effects. This part provides a series of systematic frameworks to eliminate or minimize the environmental effects on the damage detection indices for the suspension bridges.

The third part is made up of Chaps. 5 "Fatigue Monitoring of Welded Details" and 6 "Fatigue Reliability Analysis for Welded Details". Fully-welded fleet steel box girders have been widely applied in suspension bridges because of light self-weight and favorable aerodynamic performance. Long-term stress monitoring provides more accurate fatigue investigation methodology for the welded details of in-service steel box girders. The frameworks of long-term stress measurement interpretation and fatigue reliability assessment are presented in this part.

The fourth part is composed of Chaps. 7 "Vertical Deflection Monitoring of Main Girder" and 8 "Serviceability Assessment Based on Deflection Measurements". The profile of suspension bridge is one of the main concerns for both bridge designers and managers. In this part, an advanced deflection sensing system has been developed for in-service profile tracking of the suspension bridges. A serviceability assessment framework is presented for in-service suspension bridges by incorporation deflection measurements with the theories of extreme value and system reliability.

The last part consists of Chaps. 9 "Tension Force Monitoring of Main Cable" and 10 "WIM-based Fatigue Performance Investigation of Hangers". Main cables and hangers, which form the cable system, are the most critical members in the suspension bridges. Tension force of the main cable and fatigue performance of the hangers are the key concerns for the load-carrying capacity of suspension bridges. In this part, the safety assessment frameworks for main cables and hanger are presented by using the long-term measurements.

Beijing, China

Yang Deng
Aiqun Li

Acknowledgements

The authors are deeply grateful to the research staff and students in our research group for their hard work. We also express our thanks to Dr. Youliang Ding, Dr. Suren Chen, Dr. Dongming Feng, Dr. Tong Guo, and Dr. Da Wang for their invaluable advice throughout this research.

The financial supports from Beijing Scholar plan, Supporting Plan for Cultivating High Level Teachers in Colleges and Universities in Beijing with Grant IDHT20170508, National Key Technology Research and Development Program of the Ministry of Science and Technology of China with Grant 2017YFC0703602, the NSF of China with Grant 51878027, 51438002 and 51308073, Beijing Advanced Innovation Center for Future Urban Design, and Beijing Key Laboratory of Functional Materials for Building Structure and Environment Remediation are greatly appreciated.

Contents

Part 1

Chapter 1
Temperature Action Monitoring of Main Girder

1.1 Introduction

Since the first adoption on the Severn Bridge, streamlined flat steel box girders have been widely applied in long-span bridge designs because of light self-weight and favorable aerodynamic characteristics [1–3]. Currently, the maximum main spans of suspension and cable-stayed bridges with streamlined flat steel box girders have reached 1650 m (the Xihoumen Bridge, width-to-depth ratio = 11.0) and 1104 m (the Russky Island Bridge, width-to-depth ratio = 9.2), respectively [4, 5]. Long-span steel box girder bridges bring challenges to both designers and constructors, such as aerodynamic stability, fatigue crack, and construction control [3, 6, 7]. Recent field investigations on bridges with flat steel box girders suggested that temperature distribution should receive more attention [8–12]. During the past years, there have been a few investigations about the effects of temperature, such as those on fatigue damage [8], modal frequencies [10], vertical deflection [12], and displacements of expansion joints [9–11]. All these studies suggested that the temperature effect on bridges with flat steel box girders is important and complex. A comprehensive understanding of temperature characteristics of flat steel box girders is of great importance to not only new bridge design and construction but also lifetime performance evaluation and maintenance of existing bridges.

Current bridge design specifications around the world provide little information about temperature actions on flat steel box girders. British Standard Institution (BSI) [13] specified the vertical temperature difference of trapezoidal steel box girders, which is known to be very different from that of flat steel box girders. Due to larger width-to-depth ratio, the transverse temperature difference of a flat steel box girder is usually more remarkable than that of trapezoidal one and the temperature gradients in both vertical and transverse directions are needed. Some studies have identified related limitations in current design specifications. Zhang et al. [14] found that the vertical temperature difference was larger than the values provided by BSI [13] for a flat steel box girder without pavement (width-to-depth ratio = 11.8) through

Y. Deng and A. Li, *Structural Health Monitoring for Suspension Bridges*,
https://doi.org/10.1007/978-981-13-3347-7_1

heat-transfer analysis. Kim et al. [15] also used heat-transfer analysis to investigate the temperature distribution of a flat steel box girder (width-to-depth ratio = 11) during construction. Despite its importance, transverse temperature difference was not covered in existing studies due to the limited available data.

Besides, the bridge designers typically prefer that when the girders finally join together during construction stage, the actual effective temperature of the girder can be as close as possible to the reference effective temperature specified during the design stage. However, in the construction stage, it is very hard to obtain reliable information about the temperature spatial distribution information around the cross section and in turn the actual effective temperature of the girder due to the lack of sensors. Hence, ambient air temperature was often used instead in engineering practice to be checked against the reference effective temperature to decide the best time to join the girders. Apparently, ambient air temperature, despite its convenience to be measured, is different from the actual effective temperature. Using ambient air temperature as the actual effective temperature certainly causes discrepancy between the actual bridge and the original design. It thus becomes critical to appropriately transform an easily accessible ambient temperature during construction to an accurate effective temperature for deciding the best time of the final girder assembly. Chang and Im [16] presented a linear correlation model of effective temperature and ambient air temperature for a composite box-girder bridge in Republic of Korea. For flat streamlined steel box girders, the studies on relationship of ambient temperature and effective temperature have been rarely reported.

During the past decade, development of structural health monitoring system (SHMS) offers a great opportunity to provide valuable long-term data to enable an in-depth study of flat steel box girder's temperature distribution. Lucas et al. [17] statistically analyzed the effective temperature and thermal gradient of a steel box girder bridge based on 3-year monitoring data and the transverse temperature difference was not yet covered. Ding et al. [18] and Ding and Wang [19] investigated the transverse temperature differences but the vertical temperature gradient was not included due to the lack of sensors. Zhou et al. [20] investigated the vertical and transversal temperature differences of a steel box-girder suspension bridge and both measured and numerical results showed that the transverse temperature variation across the streamlined girder was significant. Ni et al. [21] also used 1-year (Year 1999) temperature data of the Ting Kau Bridge to predict the maximum and minimum effective temperature. As summarized above, SHMS-based approach opens a new door to understanding the mechanism, predicting possible damages of existing structures, and further improving the design guidelines of new structures.

This chapter proposes a comprehensive approach to study the temperature actions on flat steel box girders of long-span bridges with SHMS techniques and includes strategically placing sensors, collecting, and processing the SHMS data, and gaining comprehensive understanding of the temperature distribution on flat steel box girders. Specifically, the SHMS data is used to (1) establish a mapping relation between effective temperature with ambient air temperature, offering a convenient way to provide key information for joining bridge girders during the construction stage; (2) characterize time-dependent cyclic variation of effective temperature of bridge girders for performance assessment of some critical components of a long-span bridge; and (3) apply extreme value analysis to provide characteristic values of temperature actions including effective temperature and temperature gradients in both vertical and transverse directions.

1.2 The NSB Description and Instrumentation

The Nanxi Suspension Bridge (NSB), which was open to traffic in 2012, is in the Sichuan province of China with a main span of 820 m. The main girder of the NSB is a streamlined flat steel box girder with the dimensions shown in Fig. 1.1: the girder width is 29.7 m and the height is 3 m (width-to-depth ratio is nearly 10). The widths of the top and bottom plates are 28.1 and 22.7 m, respectively, and the thickness of the pavement is 67 mm. It is worth to mention that this type of flat box girders is popular on long-span bridges and an incomplete list of bridges using similar girders includes Zhanjiang Bay Bridge [3], Runyang Suspension Bridge [7], Jiangyin Bridge [22], and Sutong Bridge [23]. Therefore, the technology as proposed and demonstrated on this bridge and the detailed observations are representative for many long-span bridges using flat box girders. An SHMS was installed on the prototype bridge and the bridge has been continuously monitored since 2013 [12]. The temperature monitoring system was designed to measure the temperature distribution. The sensor placing strategy is as follows: (1) only the middle section of the main span was selected to place the sensors as some previous studies have revealed that the difference of temperature distributions along the span is negligible [18]; (2) 15 Pt–resistance temperature sensors were installed on the inner surface of the middle section (Fig. 1.1). The positions of sensors were arranged strategically to cover the temperature distribution around the cross section to the greatest extent. 5 sensors (ST1 to ST5) were placed on the top plate and another 5 sensors (ST10 to ST14) were placed on the bottom plate. 1.5 sensors (ST6 to ST9 and ST15) were placed on the web plate. The arrangement can allow comprehensive coverage of the temperature distribution in both transverse and vertical directions; (3) To link the ambient air temperature and the structural temperature of flat steel box girder, a temperature sensor was placed on the deck of the middle section to measure the ambient air temperature. Note that ST3 is under the middle barrier and the locations of ST1 and ST5 are near the side barriers. The barriers above the pavement were designed to be hollow without obstructing the solar radiation to the pavement.

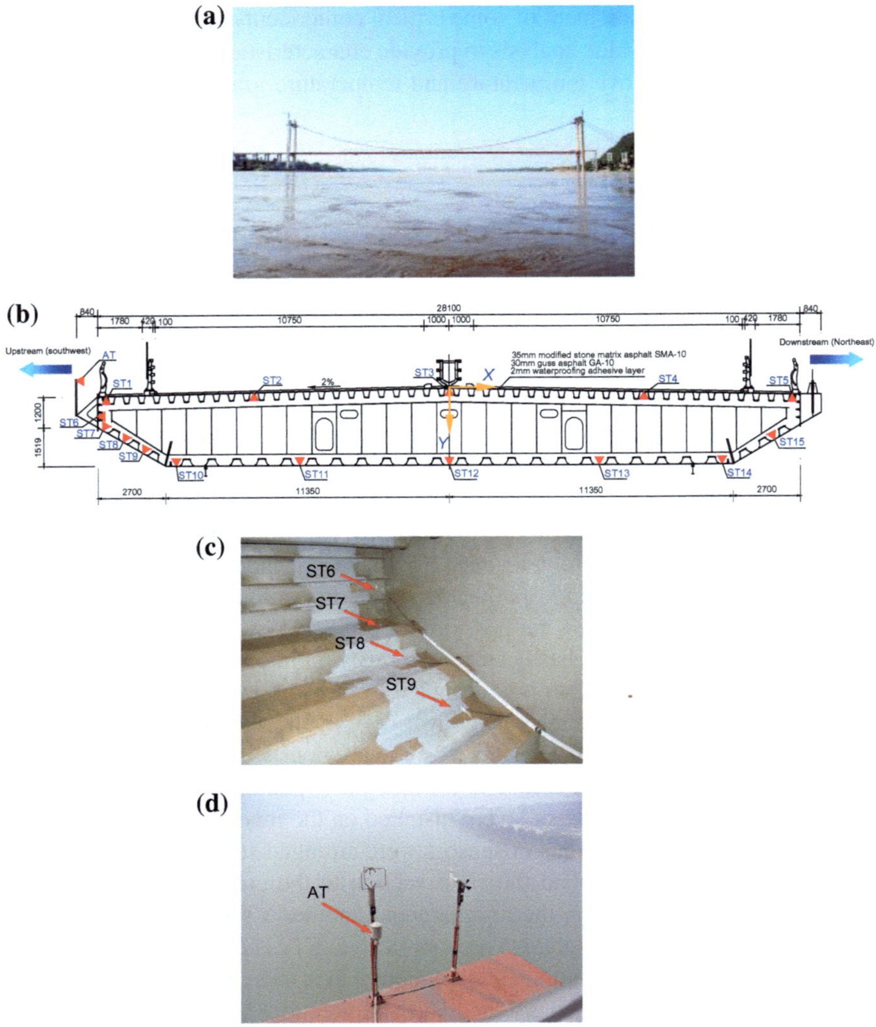

Fig. 1.1 The NSB and its temperature monitoring system: **a** the Nanxi suspension bridge, **b** temperature sensor placement in the middle cross section (unit: mm), **c** sensors in the steel box girder and **d** ambient air temperature sensor

1.3 Spatial and Temporal Characteristics of Temperature Measurements

The temperature data measured during 2015 are used. There are data sets of 240 days, which are evenly distributed over 12 months in the year 2015. The sampling frequency of all the temperature sensors is 0.0167 Hz (the period is 1 min). Since the structural temperature and ambient air temperature change at a slow rate, the measured records are averaged every 10 min. Thus, there are 144 data points each day and the size of the total 240-day data is reduced to 34,560.

Statistics are made on the raw monitoring data to provide some insightful observations on the temporal and spatial characteristics. The two days with the highest and lowest ambient temperatures of the year respectively are selected. Figure 1.2 gives the temporal-and-spatial temperature distributions. Figure 1.3 gives the spatial temperature distribution during selected time intervals. These figures are plotted from the measurements of a limited number of sensor points so the temperature distribution of the area between two adjacent sensors is derived based on linear interpolation. Note that the coordinates in these figures have been defined previously in Fig. 1.1b.

Following observations can be made from these figures: (1) the two selected days had similar patterns of temperature distribution over time during a day. The temperatures of all sensor points reached the lowest and highest values during the time intervals of 8:00–9:00 and 17:00–18:00, respectively. During the time interval of 17:00–18:00, considerable temperature differences between the top plate and web plates were observed. However, the temperature differences on the bottom plate are very small, and the temperature distribution on the bottom plate seems to be uniform all the time; (2) From 17:00 to 18:00 among the temperatures of sensor points on top plate, ST2 and ST4 were the highest, ST1 and ST5 came the second, and ST3 was the third for both July 20 and January 10. The temperature differences over 1-h time interval were much larger on July 20 than on January 10. For both days, from 8:00 to 9:00, the temperature differences on the top plate were less than those of the time interval of 17:00–18:00. It is also found that from 8:00 to 9:00 among the temperatures of sensor points on the top plate, ST3 is the highest, ST2 and ST4 are the second, and ST1 and ST5 are the third; (3) From 17:00 to 18:00, the positive temperature differences on web plates were observed for both days of July 20 and January 10. Like that of the top plate, during this time interval, the vertical positive temperature differences of July 20 were much larger than those on January 10. From 8:00 to 9:00, the trends of vertical temperature differences on July 20 and January 10 were opposite. Since these 2 days represent typical summer and winter days, this phenomenon suggests that the temperature differences may exhibit different mechanisms on summer and winter days.

After compared with some existing studies on other types of girders, the results as discussed above show that the temperature distribution of streamlined flat steel box girder has its own characteristics and differs from those of traditional trapezoidal steel box girder [13], concrete box girders [24], concrete or steel girders with GFRP panels [25, 26], and concrete-steel composite girders [27].

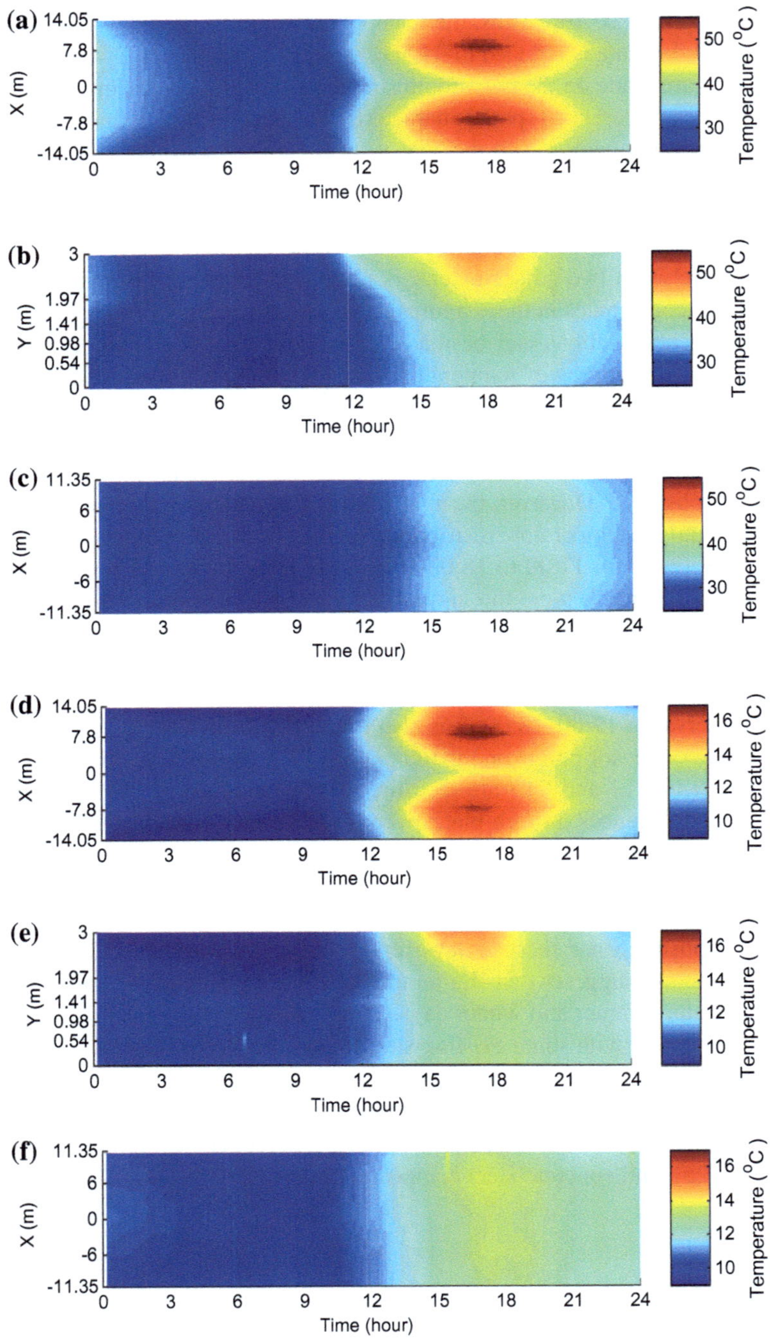

Fig. 1.2 Temporal-and-spatial distribution temperature of July 20, 2015: **a** top plate, **b** web plates, **c** bottom plate, **d** top plate, **e** web plates, and **f** bottom plate

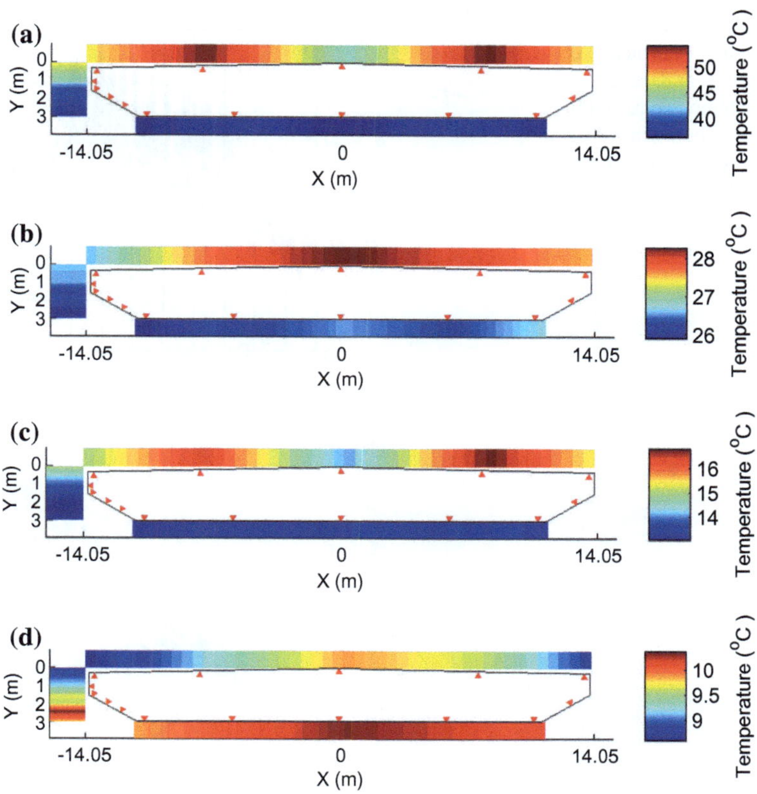

Fig. 1.3 Spatial temperature distribution in cross section of July 20, 2015: **a** 17:30–17:40, **b** 8:30–8:40, **c** 17:00–17:10, and **d** 8:00–8:10

1.4 Effective Temperature Analysis

1.4.1 Correlation of Ambient Air Temperature and Effective Temperature

The temperature spatial distribution on a cross section is typically characterized by effective temperature and temperature difference. Effective temperature provides essential information for long-span bridge designs following the design specifications. Figure 1.3 indicates that the temperature distribution on the cross section of flat steel box girder is nonuniform. Hence, the effective temperature of the whole section rather than individual temperature data measured by any single sensor should be used in the prediction of temperature effect on expansion joints and bearings. The effective temperature is the weighted mean value of the temperature at the section as defined by Tong et al. [28].

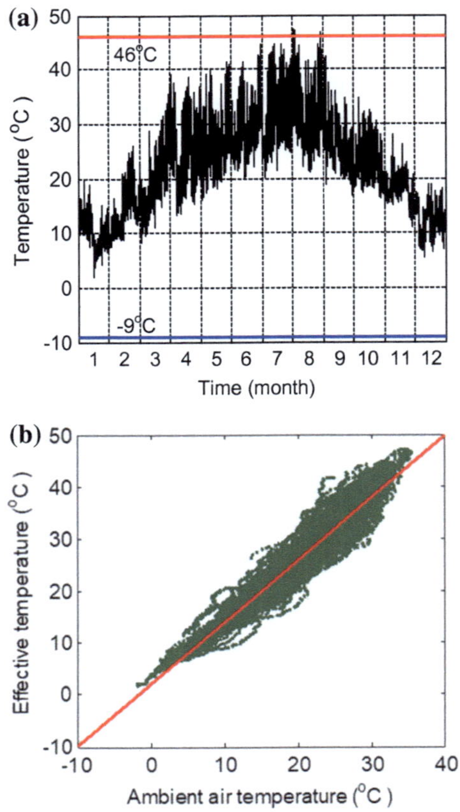

Fig. 1.4 Effective temperature of flat steel box girder: **a** 1-year data of effective temperature and **b** relationship between ambient air temperature and effective temperature

$$T_e = \frac{\int E\alpha T \mathrm{d}A}{\int E\alpha \mathrm{d}A} \tag{1.1}$$

where E and α are the Young's modulus and coefficient of the thermal expansion, respectively. T is the temperature distribution function of the cross section. dA is the differential cross-sectional area, and the integration is carried out over the whole cross section.

In this chapter, the effective temperature is approximately calculated from the measurements of 15 sensors by dividing the entire cross section into several parts. The effective temperature is calculated and illustrated in Fig. 1.4a, in which approximately half-wave annual variation is observed. The highest and lowest effective temperatures in the year 2015 were 47.4 and 2.0 °C, respectively. The Chinese bridge design code specifies the characteristic values of bridge structure's effective temperature by separating the country into three regions, which are extremely cold region, cold region, and warm region [29]. Because the location of the NSB is in the warm region of China the highest and lowest characteristic values of effective temperature are 46 and −9 °C, respectively, which are also shown in Fig. 1.4a.

In design of the NSB, the reference effective temperature was 20 °C, which was used to determine the positive and negative differences of effective temperature of the bridge structure. According to the Chinese bridge design specification, the positive and negative differences are 26 °C (46 °C − 20 °C) and −29 °C (−9 °C − 20 °C), respectively [29]. The absolute values of the positive and negative differences are very close to each other. It implies that the bridge structures are expected to have approximately equivalent temperature effects induced by positive and negative differences of effective temperature.

The effective temperature is also found to linearly correlate with the ambient air temperature as shown in Fig. 1.4b. The correlation coefficient of T_e and T_{amb} is 0.974. The proposed formula of T_e and T_{amb} of the NSB' steel box girder is

$$T_e = 1.2 \times T_{amb} + 2.2\ ^\circ\mathrm{C} \tag{1.2}$$

Substituting the reference effective temperature 20 °C into Eq. (1.2) would give the targeted ambient air temperature as 14.8 °C. Obviously, the considerable difference between the reference effective temperature and the corresponding ambient air temperature suggests that it may not be appropriate to use the ambient temperature directly in the construction control practices such as joining flat streamlined steel box girders.

1.4.2 Cycling Variation of Effective Temperature

In long-span bridge designs, the highest and lowest effective temperatures are often used to predict how much the bridge girder will expand and contract, and the expansion joints and bearings will be designed accordingly to accommodate such movements. It has been found that the expansion joints and bearings are required to have not only sufficient deformation capacity to accommodate the maximum expansion and contraction of the bridge, but also acceptable capacity, often quantified by so-called "cumulative displacements", to resist the cyclic movements [11, 21]. The comparison between the cumulative displacements and the design values of expansion joints based on field monitoring data were presented for the Sutong Bridge [11] and Ting Kau Bridge [21], respectively. Different from the direct displacement monitoring during the service period of a bridge focusing on the maximum displacements, cumulative displacements of expansion joints and bearings require detailed information such as time-dependent variations of effective temperature.

The cumulative variation of effective temperature of the NSB is investigated in this chapter. Taking the data of July 10, 2015, in Fig. 1.5 as an example, there are four key temperatures, which are the first temperature t_1, the lowest temperature t_2, the highest temperature t_3, and the last temperature t_4. The daily cycling variation of the day was composed of one heating process and two cooling processes also shown in Fig. 1.5. The values of these three processes were 20.2, −3.1 and −8.7 °C, respectively. Hence, the daily cycling variation of July 10, 2015, was 32 °C as shown

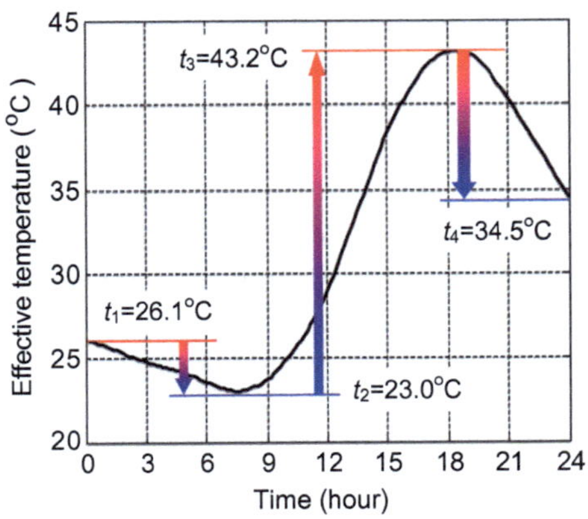

Fig. 1.5 History of effective temperature of July 10, 2015

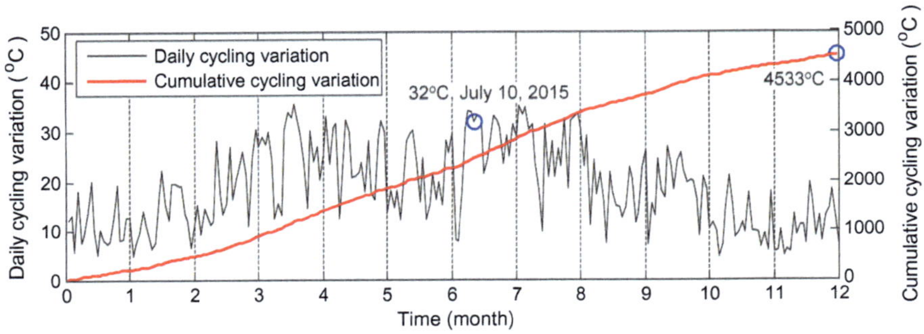

Fig. 1.6 Cycling variation of effective temperature in 2015

in Fig. 1.6. In general, it is found that there is larger daily cycling variation during summer than during winter. In addition, the cumulative cycling variation of effective temperature is also derived from the 240-day monitoring data. And 1-year cumulative cycling variation of effective temperature T_{cum} can be estimated as

$$T_{\mathrm{cum}} = \frac{365}{240} \times 4533\ ^{\circ}\mathrm{C} = 6894\ ^{\circ}\mathrm{C} \tag{1.3}$$

where 4533 °C is the cumulative cycling variation in 240 monitoring days as shown in Fig. 1.6. The number of 365 represents the total days of a year. T_{cum} can be combined with other actions to predict the cumulative displacement of expansion joints and bearings and then evaluate their wearing conditions during the service period.

1.5 Temperature Gradient Analysis

1.5.1 Transverse Temperature Differences

The difference between two sensor points i and j is defined as T_{i_j}, which equals to $T_i - T_j$ $(i, j = 1, 2, 3, \ldots, 15)$. Most design specifications typically ignore the transverse temperature differences on bridge cross sections [13, 29, 30]. However, some recent field investigations found that the transverse temperature differences on the decks of flat steel box girder [18] and concrete-steel composite box girder [27] were substantial and should be considered.

Figure 1.7 presents the time histories of transverse temperature differences on both top. Table 1.1 lists the maximum and minimum transverse temperature differences. On the top plate, the transverse temperature difference is considerable and the maximum and minimum values exceed 15 and −4 °C, respectively. In most cases, the temperatures of ST1, ST2, ST4, and ST5 are higher than that of ST3. The temperature difference T_{2_3} is larger than T_{1_3} and similarly, T_{4_3} is also larger than T_{5_3}. It can be observed from Table 1.1 that the temperature is symmetrically distributed about the center of ST3. Hence, the temperature distribution of the top plate can be captured based on the temperature data of ST1, ST2, and ST3. On the bottom plate, the maximum transverse temperature difference is only 2.2 °C for 1 year. The transverse temperature differences on the bottom plate are very small and can be neglected. Note that the temperature of ST12 will be adopted to represent the temperature of the bottom plate to assess the vertical temperature differences.

1.5.2 Vertical Temperature Differences

The temperature differences T_{3_12} between the top and bottom plates are shown in Fig. 1.8a. In the figure, it is found that the vertical temperature differences during summer are apparently higher than those during winter. Table 1.1 also gives the maximum and minimum vertical temperature differences. The data reveals that the positive temperature differences along the depth are significant while the negative differences are small.

1.5.3 Correlation Analysis of Temperature Differences

The correlation between temperature differences will be evaluated. Taking the top plate as an example, the temperature distribution seems to be easily determined since it is only related to ST1, ST2, and ST3. The temperature distribution of the top plate should be determined under 3 different comparison scenarios, which are described as T_{1_2} versus T_{1_3}, T_{2_1} versus T_{2_3} and T_{3_1} versus T_{3_2}. Within 1 year, both positive

Table 1.1 Temperature differences in cross section

	Top plate						Bottom plate				Vertical direction					
	T_{1_3}	T_{2_3}	T_{4_3}	T_{5_3}	T_{2_1}	T_{4_5}	T_{10_12}	T_{11_12}	T_{13_12}	T_{14_12}	T_{3_12}	T_{6_12}	T_{7_12}	T_{8_12}	T_{9_12}	T_{15_12}
Max	12.5	15.1	14.6	10.4	13	12.3	1.1	1.9	2.2	1.2	10.4	9.2	3.5	3.2	1.5	2.8
Min	−4.3	−2.5	−1.9	−4.5	−6.2	−1.2	−1.9	−1.3	−1.2	−1.5	−1.8	−1.9	−1.5	−1.4	−1.8	−1.5

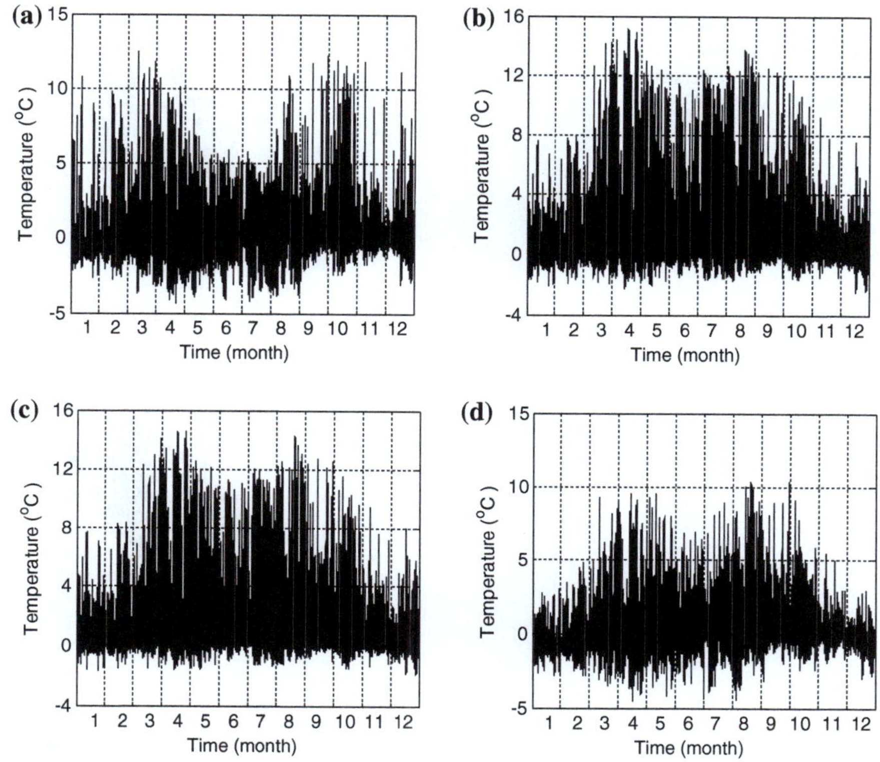

Fig. 1.7 Transverse temperature differences in top plate: **a** T_{1_3}, **b** T_{2_3}, **c** T_{4_3} and **d** T_{5_3}

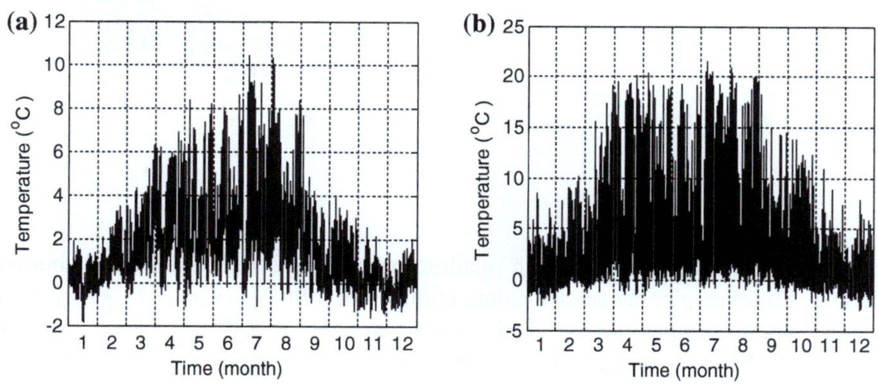

Fig. 1.8 Vertical temperature differences: **a** T_{3_12} and **b** T_{2_12}

and negative values of T_{1_2} and T_{1_3} were observed, and there are four patterns about the relative temperature distributions. The scatter diagrams of T_{1_3} versus T_{1_2}, T_{2_3} versus T_{2_1} and T_{3_2} versus T_{3_1} are presented in Fig. 1.9. In each scatter diagram,

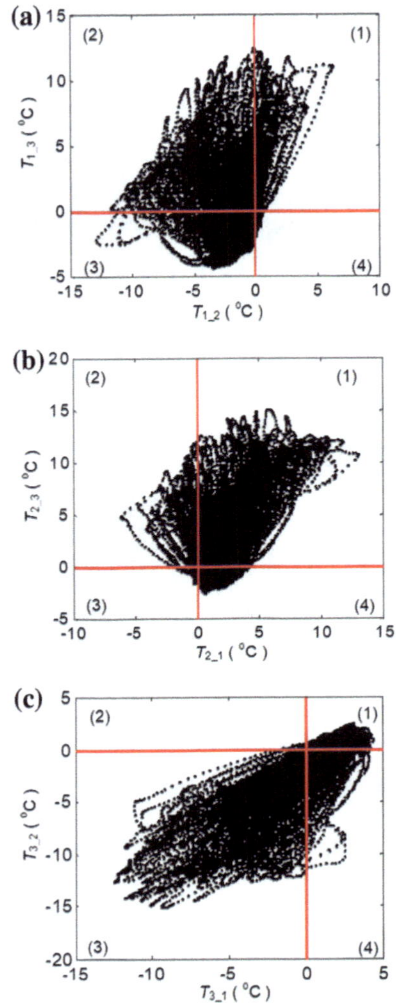

Fig. 1.9 Correlations of transverse temperature differences: **a** T_{1_3} versus T_{1_2}, **b** T_{2_3} versus T_{2_1} and **c** T_{3_2} versus T_{3_1}

there are four quadrants and each quadrant represents a temperature distribution pattern. For example, the scatter points in quadrant 1 of T_{1_3} versus T_{1_2} explain a pattern that the temperature of ST1 is higher than those of ST2 and ST3.

The number of scatter points in every quadrant is counted and the proportion is calculated by dividing the number in every quadrant by the total number of points: 34,560. Table 1.2 lists the numbers and proportions of scatter points. For the comparison between T_{1_3} and T_{1_2}, 91.5% (35.2% + 56.3%) of scatter points are found in quadrants 2 and 3 and only 8.5% (6.9% + 1.6%) of scatter points are in quadrant 1 and 4. It implies that the dominant pattern is that the temperature of ST1 is lower than that of ST2 while no clear trend can be identified between the temperatures of

Table 1.2 Data points of temperature differences of top plate in 4 quadrants

Quadrant	T_{1_3} versus T_{1_2}	T_{2_3} versus T_{2_1}	T_{3_2} versus T_{3_1}	T_{3_12} versus T_{6_12}	T_{6_12} versus T_{7_12}	T_{7_12} versus T_{8_12}	T_{8_12} versus T_{9_12}
1	2370 (6.9%)	**17,765 (51.4%)**	**14,685 (42.5%)**	**22,343 (64.6%)**	**15,089 (43.7%)**	**14,556 (42.1%)**	**10,504 (30.4%)**
2	**12,176 (35.2%)**	2226 (6.4%)	71 (0.2%)	5397 (15.6%)	7992 (23.1%)	1025 (3.0%)	4784 (13.8%)
3	**19,453 (56.3)**	765 (2.2%)	**14,467 (41.9%)**	**6144 (17.8%)**	**11,050 (32.0%)**	**18,511 (53.6%)**	**19,212 (55.6%)**
4	561 (1.6%)	**13,804 (40.0%)**	5337 (15.4%)	676 (2.0%)	429 (1.2%)	468 (1.4%)	60 (0.2%)

ST1 and ST3. Hence, the pattern of $T_{1\_2} < 0$ and $T_{1\_3} > 0$ and the pattern of $T_{1\_2} \le 0$ and $T_{1\_3} < 0$ will be the main ones to be considered.

The comparison of T_{2_3} versus T_{2_1} exhibits different trend from that of T_{1_3} versus T_{1_2}: 91.4% (51.4% + 40.0%) scatter points are collectively distributed in quadrants 1 and 4. This indicates that there is a dominant pattern that the temperature of ST2 is generally higher than that of ST1 while the relation between the temperatures of ST2 and ST3 is uncertain. Hence, in the second comparison (i.e., T_{2_3} vs. T_{2_1}), the pattern of $T_{2\_1} \ge 0$ and $T_{2\_3} \ge 0$ and the pattern of $T_{2\_1} > 0$ and $T_{2\_3} < 0$ will be considered. A similar analysis of the third comparison (i.e., T_{3_1} vs. $T_{3\_2)}$ leads to the identification of the dominant pattern of $T_{3\_1} \ge 0$ and $T_{3\_2} \ge 0$ and $T_{3\_1} \le 0$ and $T_{3\_2} < 0$. In summary, six dominant patterns, which are highlighted in Table 1.2, will be further investigated to develop the temperature gradients of the top plates.

To investigate the vertical temperature gradient on the flat steel box, the vertical temperature differences T_{3_12}, T_{6_12}, T_{7_12}, T_{8_12}, and T_{9_12} are discussed. Note that the vertical temperature differences T_{3_12}, T_{6_12}, T_{7_12}, T_{8_12}, and T_{9_12} are expected to be either all positive or all negative. It implies that there will be only two major patterns of the vertical temperature gradient, which will make the bridge design more convenient. The correlation of two nearby temperature differences is shown in Fig. 1.10. Similarly, each scatter diagram in Fig. 1.10 is divided into four quadrants. Then the number and proportion of scatter points in every quadrant are calculated and shown in Table 1.2. It can be found that the numbers of points in quadrants 1 and 3 of all scatter diagrams are the two highest ones among all: 82.4% (64.6% + 17.8%), 75.7% (43.7% + 32.0%), 95.7% (42.1% + 53.6%), and 86.0% (30.4% + 55.6%) points are in quadrants 1 and 3 of T_{3_12} versus T_{6_12}, T_{6_12} versus T_{7_12}, T_{7_12} versus T_{8_12}, and T_{8_12} versus T_{9_12}, respectively.

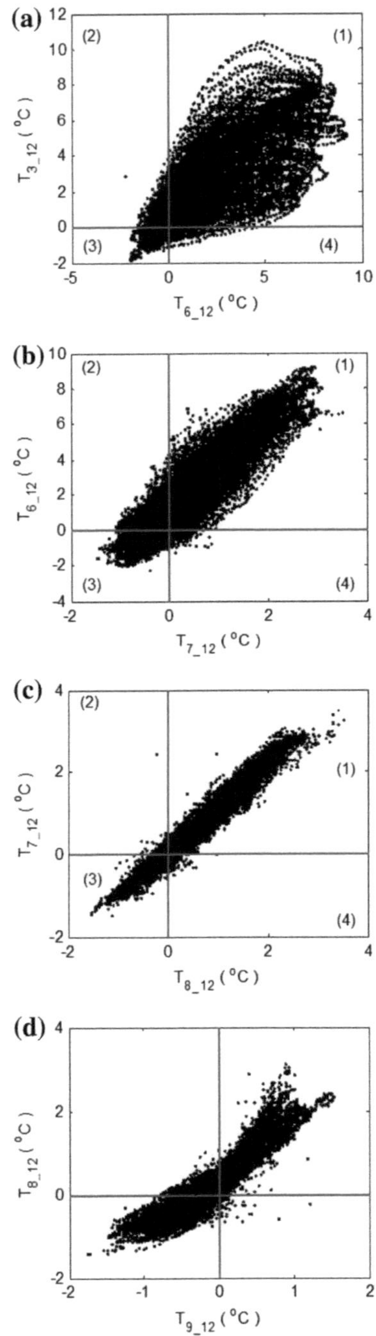

Fig. 1.10 Correlations of vertical temperature differences: **a** T_{3_12} versus T_{6_12}, **b** T_{6_12} versus T_{7_12}, **c** T_{7_12} versus T_{8_12} and **d** T_{8_12} versus T_{9_12}

1.6 Characteristic Values of Effective Temperature and Temperature Gradients

A definition of characteristic value of variable action is that an upper value with an intended probability of not being exceeded or a lower value with an intended probability of being achieved during some specific reference period [31]. According to the Chinese bridge design code [29], temperature action is a kind of variable action and the characteristic value of temperature action can be determined according to a certain quantile of maximum probability distribution during the design reference period. EN 1990 [31] specifies the return period of temperature action to be 50 years, while Chinese bridge design code specifies the design reference period to be 100 years. Hence, the exceeding probability p that the temperature action is more than its characteristic value can be expressed as

$$p = \frac{2}{100 \times v} \tag{1.4}$$

where v is the number of time block units into which a 1 year is divided. The characteristic value of temperature action T_{c} can be determined as

$$p = 1 - F_T(T_{\mathrm{c}}) = \int_{T_{\mathrm{c}}}^{+\infty} f_T(T)\mathrm{d}T \tag{1.5}$$

where $F_T(T)$ and $f_T(T)$ are the maximum probability distribution function and probability density function, respectively.

In this chapter, extreme value analysis method is adopted to estimate the characteristic values of temperature actions of the NSB's flat steel box girder. First, the daily maximum and minimum values of the ambient air temperature, the effective temperature and the transverse and vertical temperature differences are identified. Hence, the value of v in Eq. (1.4) is 365 for the 1-year data. Second, with the maximum and minimum temperature values, extreme value analysis is used to parametrically fit the data to generalized extreme value distribution (GEVD).

GEVD unites Type I, Type II, and Type III extreme value distributions into a single family to provide continuous ranges of possible shapes [12, 32]. It is parameterized with a location parameter μ, a scale parameter σ, and a shape parameter ξ. The GEVD is equivalent to Type III extreme value distribution when $\xi < 0$. The GEVD is equivalent to Type II extreme value distribution when $\xi > 0$. The GEVD becomes Type I extreme value distribution when ξ approaches 0. The GEVD distribution has a cumulative distribution function as

$$F_T(T) = \exp\left[-\left(1 + \xi\frac{T-\mu}{\sigma}\right)^{-1/\xi}\right],\ 1 + \xi\frac{T-\mu}{\sigma} > 0 \tag{1.6}$$

Table 1.3 Characteristic values of ambient air, effective temperatures, and transverse temperature differences

$AAT^{(+)}$	$AAT^{(-)}$	$ET^{(+)}$	$ET^{(-)}$	$T_{1\_3}^{(+)}$	$T_{1\_3}^{(-)}$	$T_{2\_3}^{(+)}$	$T_{2\_3}^{(-)}$	$T_{1\_2}^{(+)}$	$T_{1\_2}^{(-)}$
36.3	−15.9	49.9	−14.2	22.0	−6.9	17.7	−3.4	13.5	−23.4

Note $AAT^{(+)}$ and $ET^{(+)}$ are the highest characteristic values of ambient air temperature and effective temperature, respectively

where μ, σ, and ξ are the constants to be determined from the measurement data. The parameters are estimated by using the maximum likelihood method with 95% confidence intervals. The GEVDs fitted from the maximum and minimum temperature data are then used to extrapolate the characteristic values T_c as

$$F_T(T_c) = \exp\left[-\left(1 + \xi\frac{T_c - \mu}{\sigma}\right)^{-1/\xi}\right] = 1 - p \quad \text{for maximum characteristic value} \tag{1.7a}$$

$$F_T(T_c) = \exp\left[-\left(1 + \xi\frac{T_c - \mu}{\sigma}\right)^{-1/\xi}\right] = p \quad \text{for minimum characteristic value} \tag{1.7b}$$

Tables 1.3 and 1.4 present the characteristic values of ambient air temperature, effective temperature, and temperature differences.

1.7 Discussion of Temperature Actions for Potential Bridge Design Improvements

1.7.1 Effective Temperature

Table 1.3 presents the highest and lowest characteristic values of effective temperature, which are 49.9 and −14.2 °C, respectively. As mentioned before, Chinese bridge design code gives the highest and lowest characteristic values of effective temperature as 46 and −9 °C, respectively. Hence, those characteristic values of effective temperature as identified in this chapter are beyond the range suggested by the Chinese bridge design code [28]. It suggests that comprehensive investigations on more long-span bridges might be required to assess the applicability of the current design codes and provide useful information for possible code improvements in the future.

As discussed earlier, ambient temperature is often used during engineering practice to predict the optimal time to finally join the girders due to the convenience of being measured in the field. With multiple sensors around a cross section, more

Table 1.4 Characteristic values of vertical temperature differences

$T_{3\_12}{}^{(+)}$	$T_{3\_12}{}^{(-)}$	$T_{6\_12}{}^{(+)}$	$T_{6\_12}{}^{(-)}$	$T_{7\_12}{}^{(+)}$	$T_{7\_12}{}^{(-)}$	$T_{8\_12}{}^{(+)}$	$T_{8\_12}{}^{(-)}$	$T_{9\_12}{}^{(+)}$	$T_{9\_12}{}^{(-)}$	$T_{15\_12}{}^{(+)}$	$T_{15\_12}{}^{(-)}$
30.2	−2.3	10.1	−2.5	4.1	−1.8	3.8	−1.6	2.1	−2.0	3.3	−2.0

Note $T_{i\_j}^{(+)}$ and $T_{i\_j}^{(-)}$ are the positive and the negative characteristic values of temperature differences between sensor points *i* and *j*

accurate effective temperature data can be obtained. Equation (1.2) offers a new formula to map the relation between the effective temperature and ambient air temperature, which can be used by the bridge constructor to find out the corresponding ambient temperature to be targeted at, which can lead to the expected design effective temperature. The proposed approach, including sensor placement, can be easily applied to other bridges. With more similar studies on other bridges being conducted, a general formula will become available for this type of girders by considering uncertainties and other bridge-specific characteristics.

Equation (1.3) gave the 1-year cumulative cycling variation of effective temperature. It could be combined with the other actions to calculate the cumulative displacements of expansion joints and bearings, which can help the designer to predict the life expectancy of the existing joints and bearing and choose appropriate ones for the designs of new bridges.

1.7.2 Temperature Gradients

In this chapter, the temperature gradient of flat steel box girder is presented as two parts, which are transverse temperature differences on top plate and vertical temperature differences. First, the patterns of temperature distributions of top plate are determined by the information gathered in Tables 1.2 and 1.3. As shown in Table 1.2, there are six bold value patterns of transverse temperature differences of the top plate.

In terms of temperature distribution of the top plate as determined by T_{1_2} and T_{1_3}, there are two patterns. Pattern 1 is shown in quadrant 2 of Fig. 1.9a, where T_{1_2} is less than 0 and T_{1_3} is more than or equal to 0 are observed. In Table 1.3, T_{1_2} and T_{1_3} are shown as $T_{1\_2}^{(-)}$ and $T_{1\_3}^{(+)}$. Pattern 2 is that T_{1_2} is less than or equal to 0 and T_{1_3} is less than 0, shown as $T_{1\_2}^{(-)}$ and $T_{1\_3}^{(-)}$. There are two patterns when the temperature distribution of the top plate is determined by T_{2_1} and T_{2_3}. Pattern 3 represents T_{2_1} is more than or equal to 0 and T_{2_3} is more than or equal to 0 so T_{2_1} and T_{2_3} are equal to $-T_{1\_2}^{(-)}$ and $T_{2\_3}^{(+)}$. Pattern 4 represents T_{2_1} is more than 0 and T_{2_3} is less than 0 so they are equal to $-T_{1\_2}^{(-)}$ and $T_{2\_3}^{(-)}$.

There are also two patterns when the temperature distribution of the top plate is determined by T_{3_1} and T_{3_2}. Pattern 5 represents T_{3_1} is more than or equal to 0 and T_{3_2} is more than or equal to 0 so T_{3_1} and T_{3_2} are equal to $-T_{1\_3}^{(-)}$ and $-T_{2\_3}^{(-)}$. Pattern 6 represents T_{3_1} is less than or equal to 0 and T_{3_2} is less than 0 so they are equal to $-T_{1\_3}^{(+)}$ and $-T_{2\_3}^{(+)}$.

Making the temperature of the central sensor point ST3 to be zero, the temperatures of the other sensor points are derived and listed in Table 1.5. It can be seen in the table that Pattern 2 is very similar to Pattern 3. Pattern 4 is more unfavorable than Pattern 5 and Pattern 1 is more unfavorable than Pattern 6. Hence, the patterns of temperature difference of the top plate could be categorized into three main patterns including Pattern 1, Pattern 3, and Pattern 4. Figure 1.11 presents the profile of transverse temperature gradient of the top plate. Note that when the transverse temperature

Table 1.5 Temperature distribution in top plate

Pattern	ST1	ST2	ST3
1	22.0	45.4	0
2	-6.9	16.5	0
3	-5.7	17.7	0
4	-26.8	-3.4	0
5	-6.9	-3.4	0
6	22.0	17.7	0

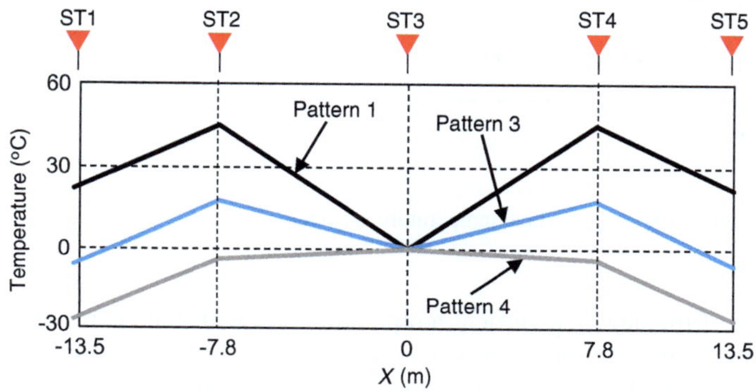

Fig. 1.11 Profile of transverse temperature gradient of the top plate

gradient is applied on the cross section, the temperature of the web and bottom plates should be set to zero and the temperature distribution of the top plate can be determined through linear interpolation.

The vertical temperature gradient is proposed using the characteristic values of vertical difference in Table 1.4. Since senor ST8 and ST15 are placed at the same height, the characteristic values of the vertical differences of these sensors are averaged. Hence, the positive and negative temperature differences of ST8 and ST15 are 3.6 and −1.8 °C, respectively. Figure 1.12 presents the profile of positive vertical temperature gradient for flat steel box girder. The positive gradient curve is divided into two parts, one of which is an exponential function, and the other is a linear function. There are four points representing ST3, ST6, ST7, and ST8 (or ST15). It is found that the exponential function can lead to the best fitting result of these four points. The positive vertical temperature gradient VTG⁺ is expressed as

$$\mathrm{VTG}^{+} = \begin{cases} T_{3_{12}}^{(+)} \times \exp(-1.12 \times Y) & 0 \leq Y \leq 2.0\ \mathrm{m} \\ 9.6 - 3.2 \times Y & 2.0\ \mathrm{m} \leq Y \leq 3.0\ \mathrm{m} \end{cases} \tag{1.8}$$

where $T_{3\_12}^{(+)}$ is equal to 30.2 °C. The vertical temperature gradient provided by BSI [13] is also plotted on the figure for comparison purposes. The BSI code specified that the positive temperature difference between the top and bottom plates was 24 °C for trapezoidal steel box girders.

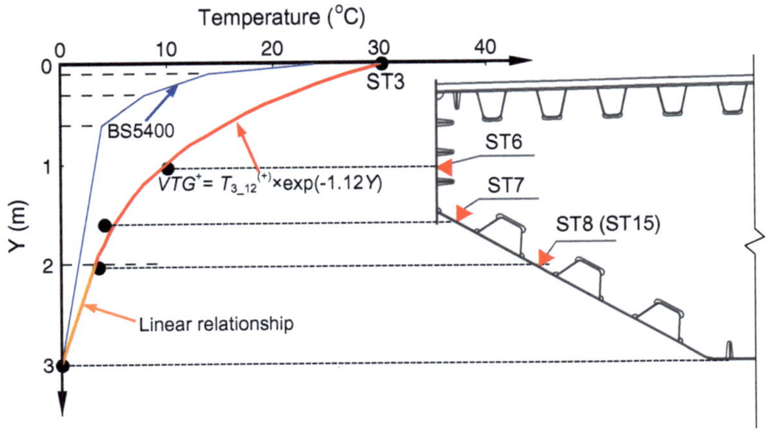

Fig. 1.12 Profile of positive vertical temperature gradient

In terms of the negative vertical temperature gradient, the characteristic values, of which the maximum absolute value is only 2.5 °C as shown in Table 1.4, are negligible. Ding et al. [18] found that the negative temperature difference between the top and bottom plates was −5.3 °C for flat steel box girder of the Runyang Suspension Bridge. Runyang Suspension Bridge is in the east of China and the width-to-depth ratio of the box girder is 12.1 (36.3 m/3 m). The negative vertical temperature gradient presented by Ding et al. [18] can provide supplemental information of temperature gradients for flat steel box girder. Note that the transverse temperature differences in Fig. 1.11 and the positive vertical temperature gradient in Fig. 1.12 should not be imposed to flat steel box girder simultaneously. If they were imposed to the girder simultaneously and Pattern 1 of transverse temperature distribution is adopted, the positive temperature difference $T_{2\_12}^{(+)}$ would be unrealistically high and reach 75.6 °C (45.5 °C + 30.2 °C).

Figure 1.8b also shows the measured T_{2_12}, of which the highest and lowest values are 21.5 and −2.9 °C, respectively. The extreme value analysis is also employed to estimate the characteristic values of T_{2_12}. The positive and negative characteristic values $T_{2\_12}^{(+)}$ and $T_{2\_12}^{(-)}$ are 30.4 and −3.7 °C, respectively. It is found in Table 1.4 that $T_{2\_12}^{(+)}$ is very close to $T_{3\_12}^{(+)}$ and hence it is unreasonable that the value of $T_{2\_12}^{(+)}$ reaches 75.6 °C. This phenomenon implies that the transverse temperature differences on top plate and the vertical temperature differences are independent from each other. They should be imposed to the cross section separately to identify the most unfavorable temperature effects.

1.8 Summary

This chapter presents a comprehensive approach to make use of long-term SHMS data to study the temperature distribution for flat steel box girders. The main contributions and findings are as follows:

1. A significant difference between ambient air temperature and effective temperature of flat steel box girder was observed. A linear mapping relation of ambient air temperature and effective temperature is presented. Such relation is of great importance to making some critical decisions dependent on temperature information during bridge construction, such as joining the flat steel box girders.
2. The cycling variation of effective temperature was investigated in this chapter. The daily effective temperature can be described as two cooling processes and one heating process. 1-year cumulative cycling variation of effective temperature was studied, which can be combined with other actions to estimate the cumulative effects of bridge structures and then predict the long-term performance of some critical components such as expansion joints and bearings.
3. The transverse temperature differences on top plate were found to be significant for flat steel box girders while those on bottom plate are negligible. To address the complexity of the temperature differences, correlation analysis was illustrated to determine the patterns of both transverse and vertical temperature difference.
4. Three patterns of transverse temperature gradient of the top plate were proposed. The positive vertical temperature differences, which are more than the values of the trapezoidal steel box girder, are characterized by using piecewise functions including an exponential function and a linear function. The results suggest that the transverse and vertical temperature differences should be separately imposed onto the cross sections to find the most unfavorable temperature-induced effects.
5. In the chapter, due to the limitation of the available data, only 1-year SHM data was used. With the future accumulation of temperature monitoring data, the characteristic values of temperature actions can be further updated by using the proposed methodology.

References

1. Brownjohn JMW, Dumanoglu AA, Severn RT, Taylor CA. Ambient vibration measurements of the Humber Suspension Bridge and comparison with calculated characteristics. Proc Inst Civ Eng. 1987;83(3):561–600.
2. Cao YH, Yim JS, Zhao Y, Wang ML. Temperature effects on cable stayed bridge using health monitoring system: a case study. Struct Health Monit. 2010;10(5):523–37.
3. Feng M. Modern bridges in China. Struct Infrastruct Eng. 2014;10(4):429–42.
4. Wang X, Song H. Zhoushan Xihoumen Bridge with the world record span length of steel box girder, China. Struct Eng Int. 2010;20(3):312–6.
5. Weber F, Distl H. Amplitude and frequency independent cable damping of Sutong Bridge and Russky Bridge by magnetorheological dampers. Struct Control Health Monit. 2014;22(2):237–54.
6. Li H, Laima S, Ou J, Zhao X, Zhou W, Yu T, Li N, Liu Z. Investigation of vortex-induced vibration of a suspension bridge with two separated steel box girders based on field measurements. Eng Struct. 2011;33(6):1894–907.

7. Deng Y, Liu Y, Feng DM, Li AQ. Investigation of fatigue performance of welded details in long-span steel bridges using long-term monitoring strain data. Struct Control Health Monit. 2015;22(11):1343–58.
8. Guo T, Li AQ, Wang H. Influence of ambient temperature on the fatigue damage of welded bridge decks. Int J Fatigue. 2008;30(6):1092–102.
9. Deng Y, Ding Y, Li A. Structural condition assessment of long-span suspension bridges using long-term monitoring data. Earthq Eng Eng Vib. 2010;9(1):123–31.
10. Ding Y, Li A. Temperature-induced variations of measured modal frequencies of steel box girder for a long-span suspension bridge. Int J Steel Struct. 2011;11(2):145–55.
11. Guo T, Liu J, Zhang YF, Pan ZH. Displacement monitoring and analysis of expansion joints of long-span steel bridges with viscous dampers. J Bridge Eng. 2015;20(9):04014099.
12. Liu Y, Deng Y, Cai CS. Deflection monitoring and assessment for a suspension bridge using a connected pipe system: a case study in China. Struct Control Health Monit. 2015;22(12):1408–25.
13. BSI. Steel, concrete and composite bridges—part 2: specification for loads: BS 5400-2:2006: 2006.
14. Zhang YP, Yang N, Li CX. Research on temperature field of steel box girder without pavement caused by the solar radiations. Eng Mech. 2011;28(6):156–62.
15. Kim SH, Park SJ, Wu J, Won JH. Temperature variation in steel box girders of cable-stayed bridges during construction. J Constr Steel Res. 2015;112:80–92.
16. Chang SP, Im CK. Thermal behaviour of composite box-girder bridges. Proc Inst Civ Eng Struct Build. 2000;140(2):117–26.
17. Lucas JM, Berred A, Louis C. Thermal actions on a steel box girder bridge. Struct Build. 2003;156(2):175–82.
18. Ding Y, Zhou G, Li A, Wang G. Thermal field characteristic analysis of steel box girder based on long-term measurement data. Int J Steel Struct. 2012;12(2):219–32.
19. Ding Y, Wang G. Estimating extreme temperature differences in steel box girder using long-term measurement data. J Central South Univ. 2013;20(9):2537–45.
20. Zhou LR, Xia Y, Brownjohn JMW, Koo KY. Temperature analysis of a long-span suspension bridge based on field monitoring and numerical simulation. J Bridge Eng. 2015;21(1):04015027.
21. Ni YQ, Hua XG, Wong KY, Ko JM. Assessment of bridge expansion joints using long-term displacement and temperature measurement. J Perform Constructed Facil. 2007;21(2):143–51.
22. Ji B, Chen D, Ma L, Jiang Z, Shi G, Lv L, Xu H, Zhang X. Research on stress spectrum of steel decks in suspension bridge considering measured traffic flow. J Perform Constructed Facil. 2012;26(1):65–75.
23. Wang H, Hu R, Xie J, Tong T, Li A. Comparative study on buffeting performance of sutong bridge based on design and measured spectrum. J Bridge Eng. 2013;18(7):587–600.
24. Hedegaard B, French C, Shield C. Investigation of thermal gradient effects in the I-35W St. Anthony Falls Bridge. J Bridge Eng. 2013; 18(9):890–900.
25. Kong B, Cai CS, Kong X. Thermal behaviors of concrete and steel bridges after slab replacements with GFRP honeycomb sandwich panels. Eng Struct. 2013;56:2041–51.
26. Kong B, Cai CS, Kong X. Thermal property analysis and applications of GFRP panels to integral abutment bridges. Eng Struct. 2014;76:1–9.
27. Zhou GD, Yi TH, Chen B, Zhang H. Analysis of three-dimensional thermal gradients for arch bridge girders using long-term monitoring data. Smart Struct Syst. 2015;15(2):469–88.
28. Tong M, Tham L, Au F. Extreme thermal loading on steel bridges in tropical region. J Bridge Eng. 2002;7(6):357–66.
29. Ministry of Transport of the People's Republic of China. General code for design highway bridges and culverts JTG D60-2004. Beijing: China Communications Press; 2004.
30. AASHTO. AASHTO LRFD bridge design specifications. 6th ed; 2012.
31. Eurocode—basis of structural design: EN 1990: 2002 + A1: 2002.
32. Coles S. An introduction to statistical modeling of extreme values. London: Springer; 2001.

Chapter 2
Bridge-Site Extreme Wind Prediction

2.1 Introduction

During the lifetime of a bridge, the mean wind velocity is one of the most important requirements for evaluating the wind resistance of the bridge. Usually, the mean wind velocity is statistically described by a random variable model. We have been particularly interested in determining the extreme velocity in a given return period using statistical extrapolation. A lot of efforts have been put into the estimation of the extreme velocity. Various probability distribution models have been used or proposed for the statistical analysis of recorded wind velocities [1–5]. The distributions most commonly used are the extreme value distributions called the Gumbel distribution, the Frechet distribution, and the Weibull distribution [6–8]. Ge and Xiang [2] predicted extreme wind velocities using these three distributions for the Yangpu Bridge in the Shanghai area. Xiao et al. [4] evaluated the capacity of the Gumbel and Weibull distributions for fitting the extreme wind speed in the Hong Kong area. After the 1970s, most researchers have deemed that the Gumbel distribution is suitable for fitting extreme wind velocity data [8–11]. As a result, the Gumbel distribution is widely adopted by structural design codes and standards throughout the world.

It should be noted that during most of the past studies on extreme wind velocity, researchers could only use wind data recorded by meteorological stations located usually far from the sites of the bridges concerned [2]. In the ideal case, there is enough wind data from the bridge site for wind-resistant design and the evaluation of wind resistance during the lifetime of the bridge. In the past decade, a significant research effort has focused on the development of structural health monitoring (SHM) for long-span bridges [12, 13]. SHMS has brought new opportunities for researchers who are interested in the wind characteristics and the wind-induced response of bridge structures [14–17]. As for the wind-resistant design of long-span bridges, the effect of mean wind velocity on structures is related to not only wind speed but also wind direction. Hence, it is valuable to study the joint distribution of wind speed and its direction using the wind data at the site of the bridge. Until now, little effort has

Y. Deng and A. Li, *Structural Health Monitoring for Suspension Bridges*,
https://doi.org/10.1007/978-981-13-3347-7_2

been put into the estimation of the joint distribution and the prediction of the extreme wind velocity using the wind data recorded by the SHM system. The most important reason is that because of the short operation time of most SHM systems, the quantity of recorded wind velocity data may be insufficient when extreme value distributions are used to estimate the extreme wind velocity of the bridges.

Hence, it is necessary to develop a new method to estimate the joint distribution model of wind speed and wind direction from a small sample measured by the SHM system of a bridge. An alternative approach for distribution fitting comes from modern information theory. The maximum entropy theory has been developed as a quantitative approach for seeking a probability distribution, among all possible distributions, that contains a minimum of spurious information [18, 19].

This chapter established a new joint distribution model of wind speed and direction based on the maximum entropy theory. A relatively small wind velocity sample, collected by the SHM system of the Runyang Suspension Bridge (RSB), is used to estimate the joint distribution model. The results of the wind velocity sample are compared with the results predicted from the proposed method and those from the Gumbel distribution. The extreme wind velocity in 100-year return period is then been predicted.

2.2 The RSB Description and Wind Monitoring Instrumentation

The prototype bridge of this chapter is the RSB shown in Fig. 2.1, which is a single-span steel suspension bridge that crosses the Yangtse River, along the highway between Zhenjiang and Yangzhou in China. The main span of the bridge is 1490 m long, making it the longest of its kind in China and the third longest in the world. The SHM system of the RSB has been established to monitor in real time the responses of the bridge under various kinds of environmental actions and mobile loads by the application of modern techniques in sensing, testing, computing, and network communication [20].

For the bridge-site wind environmental monitoring, two WA15 anemometers (Fig. 2.2) manufactured by the Vaisala Company were installed in the SHM system. One is on the top of the downstream side of the south tower (218.905 m above the ground), and the other is in the middle of the upstream side of the bridge girder (69.300 m above the ground). The anemometers work all day with a working temperature range of from −50 °C ~+55 °C and a sampling frequency of 1 Hz. The anemometers were installed to the north and define the angle of wind direction with 0° to the north, by clockwise rotation. Since May 1, 2005, the anemometers have shown their high working stability, good performance in dynamic tracking and reliability for measuring data. The anemometers record not only the wind data under normal wind conditions but also capture information during typhoons, such as "Matsa."

Fig. 2.1 View of the RSB

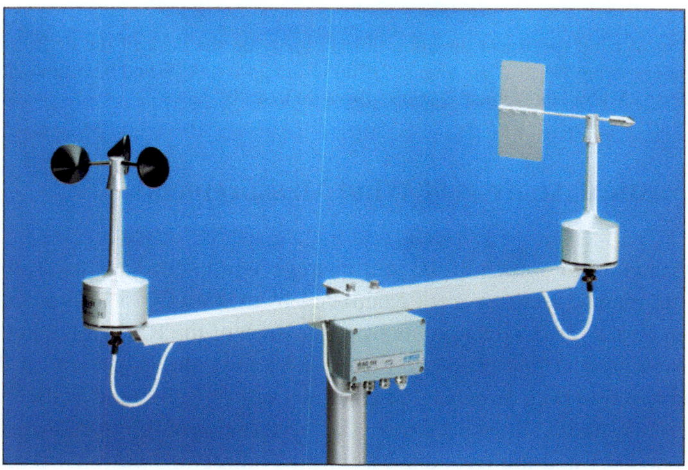

Fig. 2.2 WA15 anemometer

Typical wind speed and direction time histories recorded during typhoon "Matsa" in 2005 are shown in Fig. 2.3. Hence, the collected wind records make it possible to provide the data necessary for establishing a joint probability model between the wind speed and its direction at the site of the RSB.

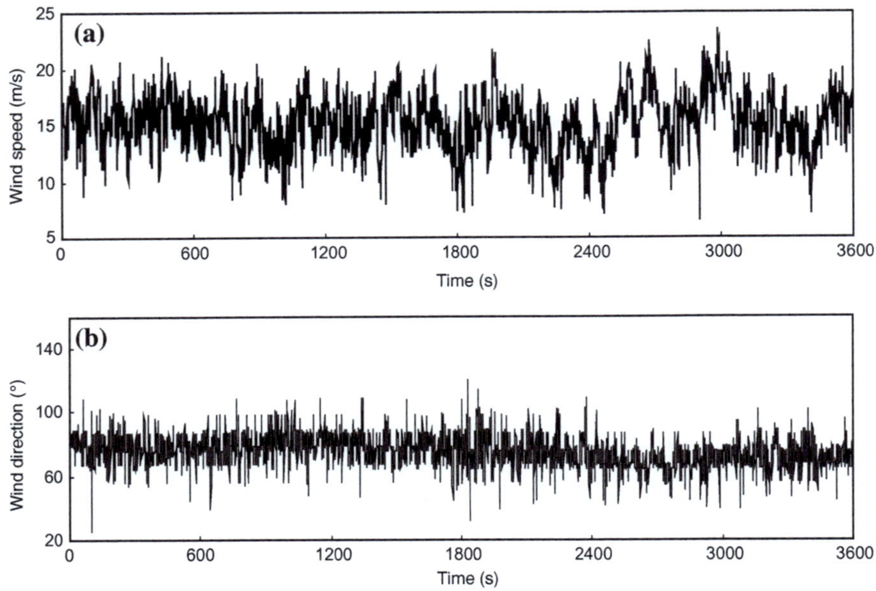

Fig. 2.3 Typical wind speed and direction time histories on the deck of the RSB: **a** wind speed history during typhoon "Matsa" on August 7, 2005 from 3:00 to 4:00 and **b** wind direction history during typhoon "Matsa" on August 7, 2005 from 3:00 to 4:00

2.3 Statistical Analysis of Wind Measurements

The daily maximum values of 10-min average wind speed for each of the eight standard compass directions were calculated from May 2005 to November 2007, resulting in a total of 543 data points from the deck of the RSB (the wind data from the deck was used in this chapter). Figure 2.4 shows the frequency of the largest daily wind speed in each direction. In the directions of E, SE, and S, the frequencies were 0.21, 0.19, and 0.18, respectively. The frequencies of the other directions were mostly around 0.1 and the frequency from the N direction was only 0.05. Because the site of the bridge is in the region of a subtropical monsoon climate influenced by the ocean monsoon blowing from the East China Sea from the E and S directions, it is logical that the frequencies of the largest daily wind speeds from the S and E were much larger than those from the other directions. Hence, Fig. 2.4 accurately reflects the wind direction characteristics at the site of the RSB. The statistical characteristics of the daily maximum wind speeds are listed in Table 2.1.

The statistical surface of the joint probability density function between the wind speed and direction was calculated. Let t_{ij} be the number of largest daily wind speeds during the wind speed interval $[U_i, U_{i+1}]$ and wind direction interval $[\theta_i, \theta_{i+1}]$. The probability of the largest daily wind speed during the above intervals was calculated approximately as t_{ij}/T, where T is the length of the sample. Considering the expres-

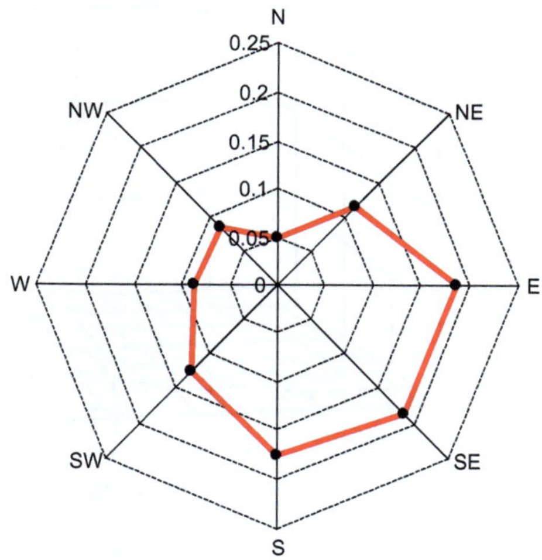

Fig. 2.4 Wind direction frequency of the largest daily wind speed

Table 2.1 Statistical characteristics of the daily maximum wind speeds (Unit: m/s)

	Wind direction								
	N	NE	E	SE	S	SW	W	NW	Non-direction
Mean value μ	6.512	8.771	7.211	7.315	7.241	7.681	8.296	7.479	7.511
Standard deviation σ	2.472	3.262	2.373	2.017	2.177	2.232	2.651	2.961	2.505

sions including $\Delta U_i = U_{i+1} - U_i$, $\Delta\theta_i = \theta_{i+1} - \theta_i$, $U_{i,m} = (U_{i+1} + U_i)/2$ and $\theta_{i,m} = (\theta_{i+1}+\theta_i)$, the joint probability density at $(U_{i,m}, \theta_{i,m})$ was calculated as

$$P_{ij} = \frac{t_{ij}}{T \cdot \Delta U_i \cdot \Delta\theta_i} \tag{2.1}$$

Using Eq. (2.1), probability density values of 22 × 8 discrete points were obtained during the wind speed interval of [0 m/s, 22 m/s] and the wind direction interval of [0°, 360°]. Figure 2.5 illustrates the statistical curved surface of the joint probability density function. It can be seen that the maximum values of the probability density were in the wind speed interval of [4 m/s, 10 m/s].

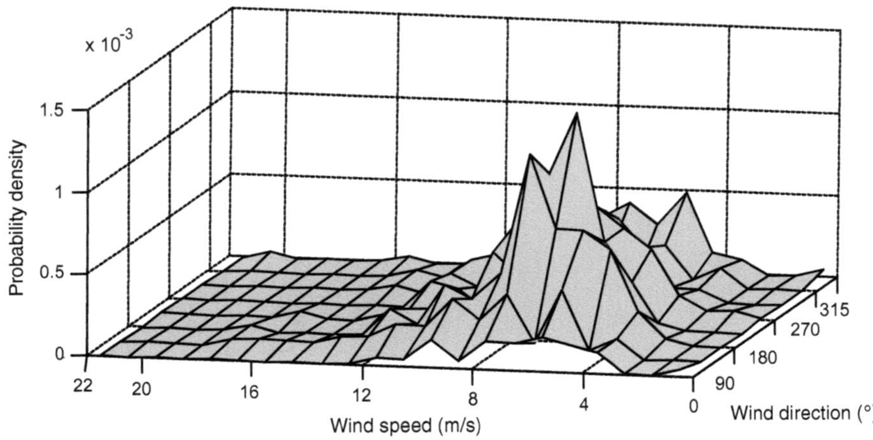

Fig. 2.5 Statistical curved surface of the joint probability density function

2.4 Maximum Entropy-Based Prediction Method

2.4.1 Basic Theory

Shannon [21] defined a measure of uncertainty, referred to as entropy. Shannon's entropy for a random event is defined as the mathematical expectation of the self-information:

$$S(x) = -\sum_{k=1}^{n} P_k(x) \ln P_k(x) \tag{2.2}$$

where $P_k(x)$ is the probability of a random variable being equal to x_k. Entropy is a positive, permutationally symmetric quantity which vanishes for a completely certain outcome and is maximum when all outcomes are equiprobable. The axiomatic characterization of entropy and its other mathematical properties are given by Kapur and Kesavan [22]. For a continuous variable x, with the density function $f(x)$, the entropy is expressed as

$$S[f(x)] = -\int f(x) \ln[f(x)] \mathrm{d}x \tag{2.3}$$

For a discrete random variable, it is

$$S[f(x_i)] = -\sum_{i=1}^{n} f(x_i) \ln[f(x_i)] \tag{2.4}$$

Jaynes [23] presented the maximum entropy principle as a rational approach for choosing a consistent probability distribution, among all possible distributions, that contains a minimum of spurious information. The principle of the maximum entropy theory states that the most unbiased estimate of a probability distribution is that which maximizes the entropy subject to constraints supplied by the available information, e.g., the moments of a random variable. The distribution so obtained is referred to as the most unbiased because its derivation involves a systematic maximization of uncertainty about the unknown information. The mathematical model of the maximum entropy theory is defined as

$$\text{MAX} \quad S = -\int f(x)\ln[f(x)]\mathrm{d}x \tag{2.5}$$

$$\text{ST} \quad \int x_i f(x)\mathrm{d}x = m_i \quad (i = 0, 1, \ldots, N) \tag{2.6}$$

where m_i is the ith order origin moment from the measured wind speed sample. N is the highest order of m_i. MAX in Eq. (2.5) means that when the entropy reaches the maximum, the best probability density function is obtained. ST in Eq. (2.6) means that when the equations are satisfied, the entropy will reach the maximum value through the adjustment of the probability density function. To account for the constraints in Eq. (2.6), the entropy function is expressed as

$$\overline{S} = S + (\lambda_0 + 1) \cdot \left[\int f(x)\mathrm{d}x - m_0\right] + \sum_{i=1}^{N} \lambda_i \cdot \left[\int x^i f(x)\mathrm{d}x - m_i\right] \tag{2.7}$$

where λ_i denotes an unknown lagrangian multiplier. To derive the function, the entropy is maximized using the usual condition:

$$\frac{\partial \overline{S}}{\partial f(x)} = 0 \tag{2.8}$$

Substituting Eq. (2.7) into Eq. (2.8) leads to the proper probability density function:

$$f(x) = \exp\left(\lambda_0 + \sum_{i=1}^{N} \lambda_0 x^i\right) \tag{2.9}$$

where N is the highest moment order of the sample. The lagrangian multipliers are determined by solving a system of nonlinear equations, Eqs. (2.6)–(2.9), by a nonlinear least square method that utilizes a Gauss Newton optimization algorithm [24]. During the calculation process, solving the system of nonlinear equations is converted to a minimum optimization problem. The optimization target function is defined as

$$T(x) = \sum_{i=0}^{N} \left[\int x^i f(x) \mathrm{d}x - m_i \right]^2 \tag{2.10}$$

It is obvious that the function of $f(x)$, among all the probable density functions with the given constraint conditions, which makes the function of $T(x)$ approximate to zero, is the best probability density function. The convergence process of the entropy to the maximum value will be demonstrated in Sect. 2.5.2.

2.4.2 Numerical Example

Because the highest order of moments in the maximum entropy method usually plays a critical role in the accuracy of the final probability density function, the fitting capacity of the method should be evaluated first. The standard normal distribution is employed to compare the fitting capacity of several maximum entropy models. One hundred sets of random numbers obeying a standard normal distribution were generated by the software of MATLAB. The length of each set was 500. The numerical integrations were carried out to determine the exceeding probability. Tables 2.2 and 2.3 list the average values of the exceeding probability and reliability indices calculated based on Eqs. (2.9) and (2.6). The reliability index of β, which is related to the exceeding probability, can be defined as

$$\beta = \Phi^{-1}[1 - P(x)] = -\Phi^{-1}[P(x)] \tag{2.11}$$

where $\Phi^{-1}(.)$ denotes the inverse standard normal cumulative distribution function.

In these tables, P_e and β_e are the theoretical values of the exceeding probability and reliability index respectively, for a given limit value. In Table 2.2, the absolute error is the absolute difference between the numerical value of the exceeding probability and P_e. In Table 2.3, the absolute error is the absolute difference between the numerical value of the reliability index and β_e. The results reveal that the in most cases the absolute errors of the fourth-order moment method are the lowest. This means that

Table 2.2 Results of the exceeding probability

Limit value	P_e	Numerical integrated value			Absolute error		
		3rd-order	4th-order	5th-order	3rd-order	4th-order	5th-order
0	0.5	0.497	0.507	0.507	0.003	0.007	0.007
1	0.1587	0.139	0.150	0.1453	0.0197	0.0087	0.0134
2	0.0228	0.0141	0.0169	0.0131	0.00869	0.00587	0.00975
3	0.0014	0.000435	0.000605	0.000243	0.000965	0.000795	0.00116
4	3.200e−05	3.544e−006	5.844e−006	6.386e−007	2.85e−05	2.62e−05	3.14e−05
5	2.910e−07	6.766e−009	1.333e−008	1.623e−010	2.84e−07	2.78e−07	2.91e−07

Table 2.3 Results of the reliability indices

Limit value	β_e	Numerical integrated value			Absolute error		
		3rd-order	4th-order	5th-order	3rd-order	4th-order	5th-order
0	0	0.00685	−0.0179	−0.0187	0.00685	0.0179	0.0187
1	1	1.0866	1.0383	1.057	0.0866	0.0383	0.057
2	2	2.194	2.122	2.225	0.194	0.122	0.225
3	3	3.329	3.237	3.488	0.329	0.237	0.488
4	4	4.491	4.383	4.843	0.491	0.383	0.843
5	5	5.679	5.562	6.287	0.679	0.562	1.287

the fourth-order moment method has the best accuracy in fitting the approximate probability density function among the three types of maximum entropy methods. Therefore, the fourth-order moment method was used to fit the joint probability density function of the wind speed and its direction.

2.5 Prediction of Extreme Wind Velocity

2.5.1 *Joint Probability Density Functions*

It is widely accepted that the effect of wind on structures is related to both the wind speed and its direction. Usually, the relationship between the wind speed and its direction is described by a wind rose diagram. For long-span bridges, it is necessary to study the joint action of wind speed and direction on the wind-induced responses of structures. Hence, the joint probability distribution model has been developed to consider the joint effect of wind speed and direction. Based on the probability density function given in Eq. (2.9), the joint probability density function is defined as

$$f(x) = l(\theta) \cdot \exp\left[\lambda_0 + \sum_{i=1}^{4} \lambda_i \left(\frac{x-u}{\sigma}\right)^i\right] \tag{2.12}$$

In Eq. (2.12), $l(\theta)$ (shown in Fig. 2.4) is the wind direction frequency function. It is obvious that the value of $l(\theta)$ is equal to 1 when the wind direction is not considered. μ is the mean value of the sample, and σ is the standard deviation. $(x-\mu)/\sigma$ is a random variable with a mean value of 0 and a standard deviation of 1. The numerical example of Wei et al. [25] shows that by using Eq. (2.12) instead of Eq. (2.8), the convergence speed can be greatly accelerated, and the difficulty of convergence failure can be reduced at the same time.

The Gumbel distribution has been the most widely used distribution for fitting extreme wind velocity data [8–11]. Thus, in this chapter, we compare the fitting results between the wind velocity data and the results predicted from our proposed method and the Gumbel distribution. The joint function based on the Gumbel distribution is defined as

$$f(x) = l(\theta) \cdot \frac{1}{a} \cdot \exp\left(-\frac{x-b}{a}\right) \cdot \exp\left[-\exp\left(-\frac{x-b}{a}\right)\right] \tag{2.13}$$

where $l(\theta)$ is the wind direction frequency function. a and b are the scale parameter and location parameter, respectively.

2.5.2 Estimation of Model Parameters

In this chapter, Eq. (2.12) was adopted rather than the original Eq. (2.9) because of the advantage of $(x - \mu)/\sigma$ in the convergence. Firstly, without considering the wind direction, the total 543 sample points were used to evaluate the convergence performance of Eqs. (2.12) and (2.9). The highest moment orders of Eqs. (2.12) and (2.9) were 4. The convergence processes are demonstrated in Figs. 2.6 and 2.7. Note that the iteration step of h in the figures is specifically the iteration step in the Gauss–Newton optimization algorithm. The target function of $T(x)$ has already been defined in Eq. (2.10).

Figure 2.6 shows that the iteration step and the target function came to the value of zero after only 23 iterations when Eq. (2.12) was involved. Figure 2.7 shows the convergence process of solving the nonlinear equation system including Eq. (2.9). The value of the target function in Fig. 2.7b was still very large after 200 iterations and it seems almost impossible to decrease to zero. Hence, we stopped the solving process. In Fig. 2.7a, a vertical dotted line divides the curve of iteration step into two phases. In phase I, (from about 0 to 80 times), a slight oscillation in the iteration process is observed. In phase II (from about 80 to 200), a severe oscillation is observed. The curve in Fig. 2.7a shows that the more the solving process iterates, the more severe the oscillation of iteration step may be. It can be concluded that Eq. (2.12) greatly

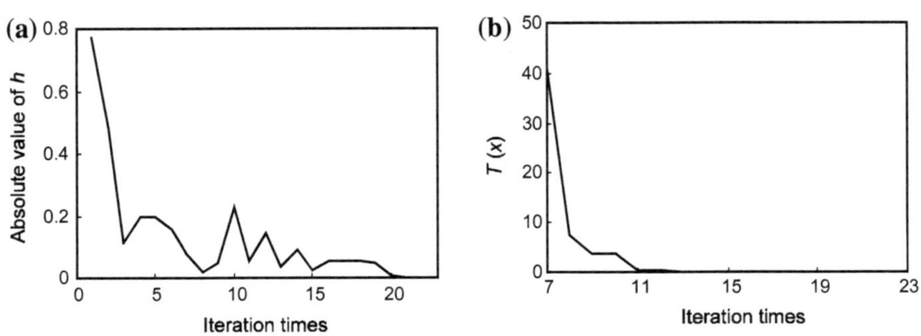

Fig. 2.6 Convergence process using Eq. (2.12): **a** iteration step of h and **b** target function of $T(x)$

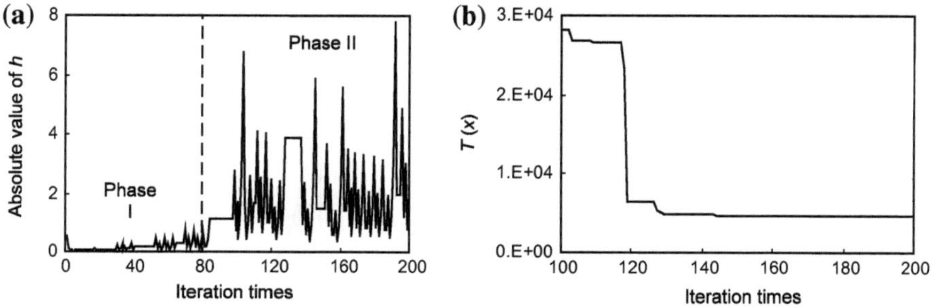

Fig. 2.7 Convergence process using Eq. (2.9): **a** iteration step of h and **b** target function of $T(x)$

Table 2.4 Parameters of the maximum entropy models

Parameters	N	NE	E	SE	S	SW	W	NW	Non-direction
λ_0	−0.892	−0.960	−0.838	−0.766	−0.913	−0.881	−0.980	−0.729	−0.842
λ_1	0.0125	−0.593	−0.423	−0.741	−0.0986	−0.630	−0.0414	−0.580	−0.384
λ_2	−0.537	−0.376	−0.608	−0.746	−0.504	−0.532	−0.351	−0.787	−0.602
λ_3	−0.0042	0.260	0.160	0.326	0.0337	0.268	0.0164	0.247	0.142
λ_4	0.0031	−0.0648	−0.0134	−0.0383	−0.0018	−0.0444	−0.0322	−0.0208	−0.0103

Table 2.5 Parameters of the Gumbel distributions

Parameter	N	NE	E	SE	S	SW	W	NW	Non-direction
Location parameter a	7.727	10.484	8.490	8.441	8.346	8.874	9.608	9.167	8.862
Scale parameter b	2.379	3.599	3.093	2.755	2.233	2.644	2.598	4.376	3.306

improves the convergence performance of the maximum entropy method compared with the original Eq. (2.9).

Table 2.4 shows the eight groups of model parameters of the joint probability models for eight compass directions using Eq. (2.12). The model parameters of Gumbel distributions are shown in Table 2.5. The model parameters without considering the wind direction are also shown in these tables.

Figure 2.8 compares the fitting results in non-direction from the proposed method with those from the Gumbel distribution. It is clear that the result from the maximum entropy theory gave a much better fit than that of the Gumbel distribution. In Fig. 2.8, the curve of the Gumbel density function offsets to the higher wind speed, so it is inevitable that the Gumbel distribution model will lead to a lower predicted value. Kolmogorov–Smirnov testing was employed to judge whether the wind speeds obeyed the Gumbel distribution. In goodness of fit tests, the total wind speed sample points were assumed to follow the Gumbel distribution with a location parameter of 8.862 and a scale parameter of 3.306. The significance level of the testing was set to

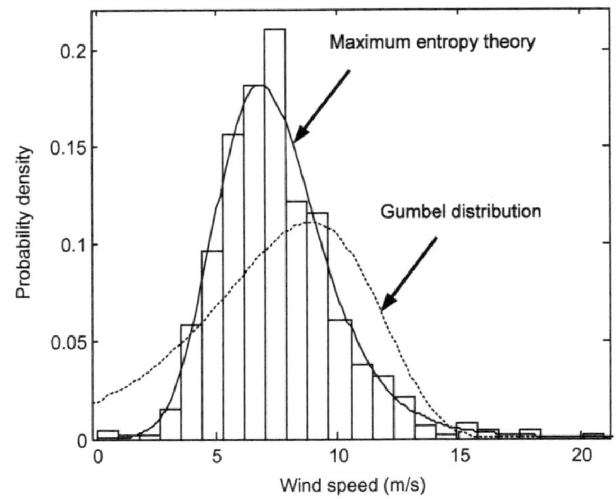

Fig. 2.8 Fitting result in non-direction

0.05. The p value of the testing, that is 2.296e−013, was far below the significance level of 0.05. The assumption of the Gumbel distribution was rejected.

Figure 2.9 shows the fitting results between the analytical expression of the probability density function and the histograms of the maximum daily wind speeds. In most wind directions, the probability density function coincides with the wind monitoring data.

2.5.3 Extreme Wind Velocities

The expected wind speed U_{100} for the 100-year return period in eight compass directions and the non-directional case at the deck level of the RSB can be obtained according to the method stated in Appendix. The predictions of extreme wind velocity are shown in Table 2.6. For comparison, the predictions of extreme wind velocity have also been calculated using the Gumbel distributions.

Table 2.6 Extreme wind velocity in 100-year return period

Model type	Wind direction								
	N	NE	E	SE	S	SW	W	NW	Non-direction
Maximum entropy model	15.7	20.1	22.7	17.6	15.7	16.5	15.8	25.9	24.2
Gumbel distribution mode	12.5	18.1	15.3	14.4	13.2	14.4	15	18.4	16.7

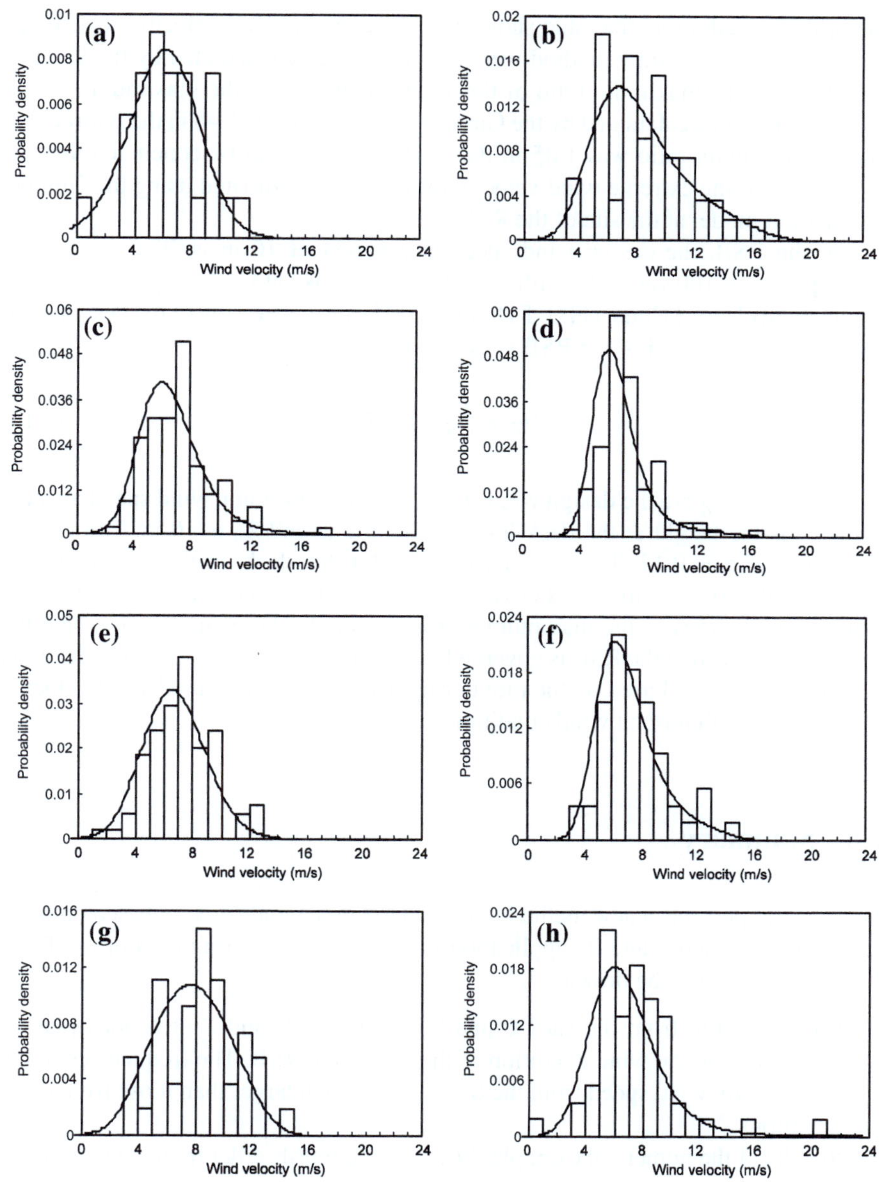

Fig. 2.9 Fitting results in eight directions: **a** N direction, **b** NE direction, **c** E direction, **d** SE direction, **e** S direction, **f** SW direction, **g** W direction and **h** NW direction

The values for the directions of N, SE, S, SW, and W were relatively smaller, but the values for the directions of NE, E, NW and non-direction were larger when calculated using the Gumbel distributions. Hence, the extreme wind speeds vary with the direction and this should be considered in the wind-induced response analysis of the bridge. In addition, the table shows that the extreme wind velocities calculated by the maximum entropy method were larger than those calculated by the Gumbel distribution. The maximum ratio of the value calculated by the maximum entropy theory to the value calculated by the Gumbel distribution was 1.48 in the E direction, and the minimum ratio was 1.05 in the W direction. It can be concluded that the predictions of the extreme wind velocity may be underestimated using the Gumbel distribution model at the site of the RSB.

For the RSB, the design wind speed at a height of 10 m is 29.1 m/s for the return period of 100 years. According to the Wind-resistant Design Specification for Highway Bridges of China (JTG-T D60-01-2004), the conversion of the design wind speed at the deck level of the RSB is expressed as

$$U_Z = \left(\frac{Z}{Z_0}\right)^{\alpha} \cdot U_0 \tag{2.14}$$

where U_0 and U_Z are the design wind speed and the conversion wind speed, respectively. Z is the height at the deck level of the bridge and Z_0 is 10 m. The ground roughness of α is 0.12. The design wind speed at the deck level U_z is calculated with the value of 36.7 m/s. It is obvious that the design wind speed is much larger than the predictions of the maximum entropy method. The main reason is that the design wind speed of the RSB is governed usually by typhoon events, while the wind velocity estimation, based on the monitoring data of daily measured wind velocity, is associated with normal wind conditions.

2.6 Summary

This chapter presents a new method to predict extreme wind velocity based on maximum entropy theory and its application to the wind monitoring data of the RSB. The conclusions can be drawn:

1. Wind blowing from the East China Sea from the E and S directions plays a critical role in the wind condition of the site of the RSB. The frequencies of the largest daily wind speed from the S and E are much larger than those from other directions.
2. Results of the numerical example show that the fourth-order moment method on the basis of maximum entropy theory has better accuracy in fitting the approximate probability density function than the third-order moment method or the fifth-order moment method.
3. The convergence performance of the maximum entropy method is greatly improved by the normalization of the original variable to a variable with a mean value of 0 and a standard deviation of 1.
4. In non-direction, the fitted curve of the Gumbel density function offsets to the higher value of wind speed and the result of Kolmogorov–Smirnov testing shows

that the wind speed sample does not obey the Gumbel distribution. But in most wind directions, results from the density function of maximum entropy method coincide with the wind monitoring data.

5. The extreme wind velocities calculated by the maximum entropy method are larger than those calculated by the Gumbel distribution. Thus, the predictions of the extreme wind velocity may be underestimated using the Gumbel distribution model at the site of the RSB. Finally, the design wind speed is much larger than the predictions of the maximum entropy method.

Appendix

Estimation of the extreme wind velocity $U_{\max}$ in the return period of T_0 years.

When the joint probability density function $f(U, \theta)$ has been obtained according to n_0 wind velocity sample points in each year, the times of sample appearance in the wind direction interval of $[\theta_j, \theta_{j+1}]$ can be calculated as follows

$$N_j = T_0 n_0 \int_{\theta_j}^{\theta_{j+1}} \mathrm{d}\theta \int_0^{\infty} f(U, \theta)\mathrm{d}U \tag{2.15}$$

During this wind direction interval, the guarantee rate which the extreme wind velocity has not exceeded can be expressed as

$$P = \left(1 - \frac{1}{N_j}\right) \int_{\theta_j}^{\theta_{j+1}} \mathrm{d}\theta \int_0^{\infty} f(U, \theta)\mathrm{d}U \tag{2.16}$$

Substituting Eq. (2.15) into Eq. (2.16), the guarantee rate P can be expressed as

$$P = \int_{\theta_j}^{\theta_{j+1}} \mathrm{d}\theta \int_0^{\infty} f(U, \theta)\mathrm{d}U - \frac{1}{T_0 n_0} \tag{2.17}$$

Meanwhile, the guarantee rate P also equals the volume enclosed by the curved surface of the joint probability function $f(U, \theta)$ and the rectangular region $0 \le U \le U_{\max}$, $\theta_j \le \theta \le \theta_{j+1}$, which can be expressed as

$$P = \int_{\theta_j}^{\theta_{j+1}} \mathrm{d}\theta \int_0^{U_{\max}} f(U, \theta)\mathrm{d}U \tag{2.18}$$

Combining Eq. (2.17) with Eq. (2.18), the following equation can be obtained:

$$\int_{\theta_j}^{\theta_{j+1}} \mathrm{d}\theta \int_{U\mathrm{max}}^{\infty} f(U,\theta)\mathrm{d}U = \frac{1}{T_0 n_0} \tag{2.19}$$

According to Eq. (2.19), extreme wind velocity $U_{\max}$ within T_0-year the return period can be obtained.

References

1. Simiu E. Modern developments in wind engineering: part 1. Eng Struct. 1981;3(4):233–41.
2. Ge YJ, Xiang HF. Statistical study for mean wind velocity in Shanghai area. J Wind Eng Ind Aerodyn. 2002;90(12–15):1585–99.
3. An Y, Pandey MD. A comparison of method of extreme wind speed estimation. J Wind Eng Ind Aerodyn. 2005;93(7):535–45.
4. Xiao YQ, Li QS, Li ZN, Chow YW, Li GQ. Probability distributions of extreme wind speed and its occurrence interval. Eng Struct. 2006;28(8):1173–81.
5. Deng Y, Ding Y, Li A. Prediction of extreme wind velocity at the site of the Runyang Suspension Bridge. J Zhejiang Univ Sci A. 2011;12(8):605–15.
6. Gumbel EJ. Statistics of extremes. New York: Columbia University Press; 1958.
7. Mayne JR. The estimation of extreme winds. J Wind Eng Ind Aerodyn. 1979;5(1–2):109–37.
8. Simiu E, Scanlan RH. Wind effects on structures. 3rd ed. New York: Wiley; 1996.
9. Holmes JD. Wind loading of structures. London: Spon Press; 2001.
10. Li GQ, Li QS. Theory and its application of time-dependent reliability of engineering structures. Beijing: Science Press; 2001.
11. Cook NJ. Confidence limits for extreme wind speeds in mixed climates. J Wind Eng Ind Aerodyn. 2004;92(1):41–51.
12. Ko JM, Ni YQ. Technology developments in structural health monitoring of large-scale bridges. Eng Struct. 2005;27(12):1715–25.
13. Deng Y, Ding YL, Li AQ. Structural condition assessment of long-span suspension bridges using long-term monitoring data. Earthq Eng Eng Vib. 2010;9(1):123–31.
14. Xu YL, Zhu LD, Wong KY, Chan KWY. Field measurement results of Tsing Ma suspension Bridge during Typhoon Victor. Struct Eng Mech. 2000;10(6):545–59.
15. Zhu LD, Xu YL, Zhang F, Xiang HF. Measurement of aerodynamic coefficients of tower components of Tsing Ma Bridge under yaw winds. Wind Struct. 2003;6(1):53–70.
16. Deng Y, Li AQ, Ding YL, Sun J. Long-term monitoring and analysis of the wind environment at the site of Runyang Suspension Bridge. Acta Aerodynamica Sinica. 2009;27(6):632–8.
17. Wang H, Li AQ, Jiao CK, Li XP. Characteristics of strong winds at the Runyang Suspension Bridge based on field tests from 2005 to 2008. J Zhejiang Univ Sci A. 2010;11(7):465–76.
18. Pandey MD. Direct estimation of quantile functions using the maximum entropy principle. Struct Saf. 2000;22(1):61–79.
19. Chen X, Qi H, Zhang Y, Wu C. Optimal design of a two-stage mounting isolation system by the maximum entropy approach. J Sound Vib. 2001;243(4):591–9.

20. Li AQ, Miao CQ, Li ZX, Han XL, Wu SD, Ji L, Yang YD. Health monitoring system for the Runyang Yangtse River Bridge. J Southeast Univ (Natural Science Edition). 2003;33(5):544–8.
21. Shannon CE. The mathematical theory of communication. Urbana: The University of Illinois Press; 1949.
22. Kapur JN, Kesavan HK. Entropy optimization principles with applications. SanDiego: Academic Press Inc.; 1992.
23. Jaynes ET. Information theory and statistical mechanics. Phys Rev. 1957;106(4):620–30.
24. Chen JH. The convergence analysis of inexact Gauss-Newton methods for nonlinear problems. Comput Optimization Appl. 2008;40(1):97–118.
25. Wei Z, Ye JH, Shen SZ. Engineering application of the maximum entropy reliability theory. J Vib Shock. 2007;26(6):146–52.

Part 2

Chapter 3
Measurement-Based Damage Detection for Expansion Joints

3.1 Introduction

The most important issues in SHM are to detect, locate, and assess the extent of damage so that its remaining life can be known and possibly extended [1]. The general methodology of structural damage detection was extracting meaningful features from the measured data [2–4]. However, it is well known that environmental conditions such as traffic loadings and environmental temperature will cause changes in the measured structural response which may mask the changes caused by structural damage. Therefore, it is of paramount importance to characterize the normal variability of structural responses due to environmental effects and discriminate such normal variability from abnormal changes in structural responses caused by structural damage [5, 6].

Considerable research efforts have been devoted to investigating the influences of environmental conditions on bridge modal frequencies [5–8]. For instance, Cornwell et al. [7] observed the variability of modal frequencies by up to 6% over a 24-h period on the Alamosa Canyon Bridge. Ding et al. [6] used improved neural networks to present the correlation model between modal frequencies of RSB and environmental factors including temperature, traffic condition, and wind speed.

It should be noted that the measured modal frequencies can only reflect the global dynamic characteristics of the bridge, which is difficult for condition assessment of local bridge components. The expansion joint, which is an important local component for a bridge, is studied in this chapter. Therefore, it needs to search a new way instead of modal frequency to detect the damage to expansion joint. It is well known that the main role of expansion joint is to meet the relative movement between main girder and abutment, thus it is very natural to link the state of expansion joint and the measured longitudinal displacement [9]. Under the operation condition, the measured displacements may have correlation in an inherent pattern with the environmental conditions including temperature, traffic, and wind. Once expansion joint is damaged, the correlation pattern may change. Then whether the expansion joints are healthy or

Y. Deng and A. Li, *Structural Health Monitoring for Suspension Bridges*,
https://doi.org/10.1007/978-981-13-3347-7_3

damaged will be decided rationally by detecting the change of the correlation pattern. Ni et al. [10] presented a procedure for condition assessment of expansion joints based on long-term monitoring data of displacement and temperature. The results revealed that the movements of expansion joints are mainly induced by temperature fluctuation. However, in their studies, the variations of expansion joint displacements induced by traffic loadings and wind speed were not considered.

The main focus of this chapter was to propose a procedure for damage detection of expansion joints of the suspension bridges under changing environmental factors including temperature, traffic condition, and wind. The study of the variations in the expansion joint displacements was presented as a function of temperature, traffic loadings, and wind speed, based on which the simple regression models were established and the dominant environmental factors were determined. The multiple linear regression models were obtained to describe the correlation between the expansion joint displacements and the dominant environmental factors. X-bar control chart, a kind of the statistical control charts, was utilized for damage detection if the future monitoring displacement data disobey the normal pattern. The feasibility of the proposed technique was demonstrated using 130 days data obtained by the structural health monitoring system (SHMS) of the Runyang Suspension Bridge (RSB).

3.2 Displacement Monitoring of the RSB

The prototype bridge of this chapter is the RSB, which have been already introduced in Chap. 2. Figure 3.1 gives the layout of the sensors on the RSB including anemometers, accelerometers, strain gages, temperature sensors, displacement sensors, etc. [11]. The displacement sensors, manufactured by ASM (a company in Germany), were used to measure the displacements of two expansion joints located at the north (Yangzhou) and south (Zhenjiang) abutments at the level of the bridge

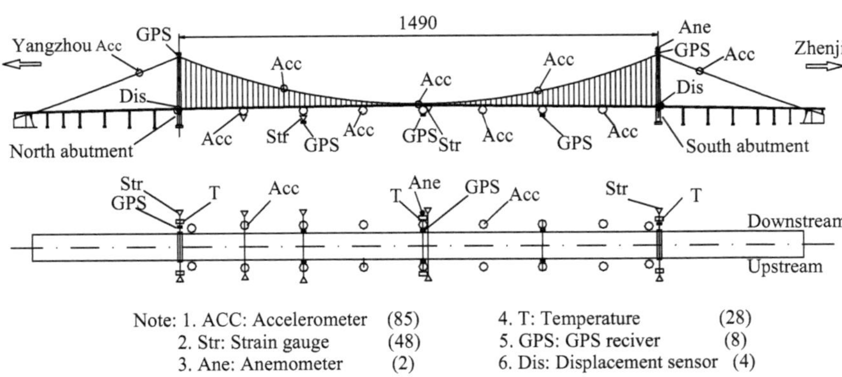

Fig. 3.1 Sensor placement of the RSB (unit: m)

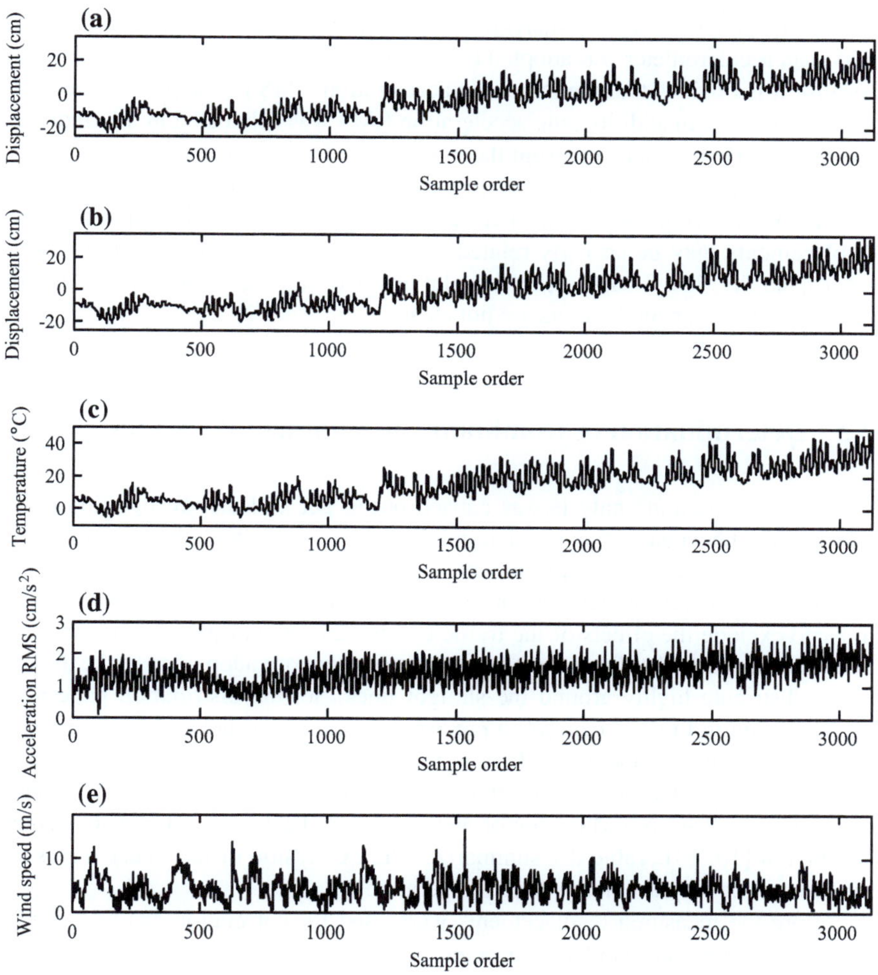

Fig. 3.2 Hourly averaged original measurements: **a** displacements of the north abutment, **b** displacements of the south abutment, **c** averaged temperature from the 28 temperature sensors, **d** RMS of the vertical acceleration response from the mid-span accelerometer, and **e** wind speeds from the mid-span anemometer

deck. The sampling frequency of the displacement sensors is 1 Hz. A total of 130-day data shown in Fig. 3.2, which covered the measurements from January to June in the year 2006, were used in this chapter.

Figure 3.2a, b show the hourly averaged displacement values of the north and south abutments, respectively. Figure 3.2c–e, in which the calculation time of each data point is also an hour, show the environmental factors including temperature, traffic and wind, respectively. The temperature data from all 28 sensors were averaged to a value as the representative temperature of the bridge. It should be noted that the root mean square (RMS for short) of the vertical acceleration responses from the mid-span accelerometer was adopted to represent the intensity of the bridge's traffic because it is obvious that the larger the acceleration RMS is, the greater the traffic intensity is [12]. In addition, the acceleration RMS was calculated in the frequency band of 0–3 Hz in order to prevent the interference from ambient noise.

In Figs. 3.2a–c, it can be easily found that the change regularities of the displacement and temperature sequences are almost the same so the expansion joint displacements may be strongly related to the environmental temperature. On the other hand, in Figs. 3.2a, b, d, e, the relations between the displacements and the other two environmental factors are not clear.

3.3 Determination of Dominant Environmental Factors

The linear regression analysis was carried out to get a clearer recognition of the relationship between the expansion joint displacements and the environmental factors. The temperature–displacement scatter diagrams are plotted in Fig. 3.3. In this figure, T represents temperature and the range of T is (−5.15, 46.75 °C). In order to exclude the effects of the traffic and wind on the displacements, the daily averaged values of the displacements and temperature are adopted to make the data points distribute highly around the straight lines. So, the distribution patterns of the data point in Fig. 3.3 can give a precise description for the correlation between displacement and temperature. In this figure, a linear increase in displacement is observed with the increase in temperature of the bridge. Furthermore, a linear regression analysis is further preformed to model the temperature–displacement correlation [10, 13]. Table 3.1 summarizes the expression of the linear regression function of the displacement versus temperature.

Before the measured displacements were used for correlation analysis of traffic–displacement and wind–displacement, the temperature effect on the measured displacements should be removed. It was achieved by normalizing all the measured displacements to a fixed reference temperature with the use of the established temperature–displacement correlation models shown in Table 3.1. The reference temperature was set to 20 °C. By presenting the reference temperature into the temperature–displacement models shown in Table 3.1, a nominal displacement d_{TR} was obtained. Likewise, by feeding the temperature measurement data into the

Table 3.1 Linear regression functions between displacements and environmental factors

	Temperature–displacement	Traffic–displacement	Wind–displacement
North	$d_M = 1.01T - 17.68$	$d^T = -0.54V + 2.88$	$d^V = -0.014W + 2.20$
South	$d_M = 1.02T - 15.50$	$d^T = -0.67V + 5.62$	$d^V = -0.007W + 4.72$

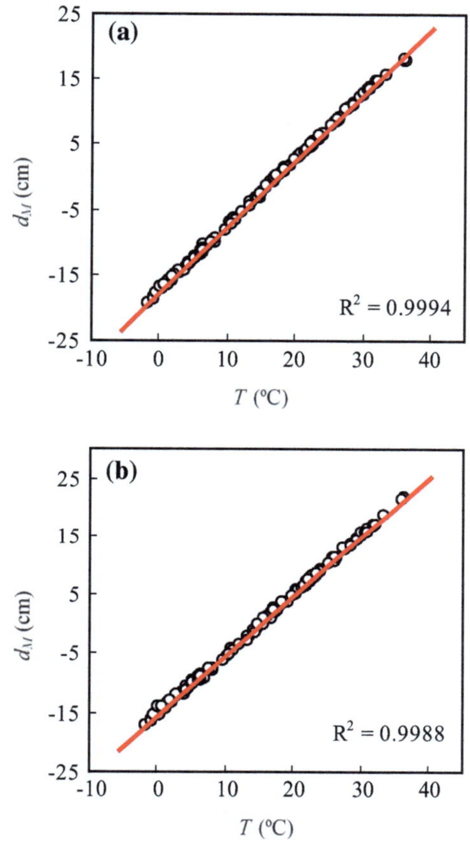

Fig. 3.3 Correlation between displacement and temperature (daily averaged): **a** north abutment and **b** south abutment

model, a temperature-induced displacement d_T was predicted. Then, the temperature-normalized displacement by removing temperature effect can be obtained by

$$d^T = d_M - (d_T - d_{TR}) \tag{3.1}$$

where d^T is the temperature-normalized displacement; d_M is the measured displacement (hourly averaged).

After removing the effects of temperature on the original displacement measurements, the correlation between the temperature-normalized displacements d^T and traffic conditions was investigated. The acceleration RMS of the mid-span accelerometer was chosen as the representative of the traffic conditions. Figure 3.4 shows the scatter diagrams of d^T versus acceleration RMS. In this figure, V represents acceleration RMS and the range of V is (0.08 , 2.70 cm/s^2) Note that, the data points in Fig. 3.4 were selected when the wind speed was less than 2 m/s, thus d^T is only affected by traffic conditions. So, the distribution patterns of the data point in Fig. 3.4 can give a reasonable description for the correlation between displacement and traf-

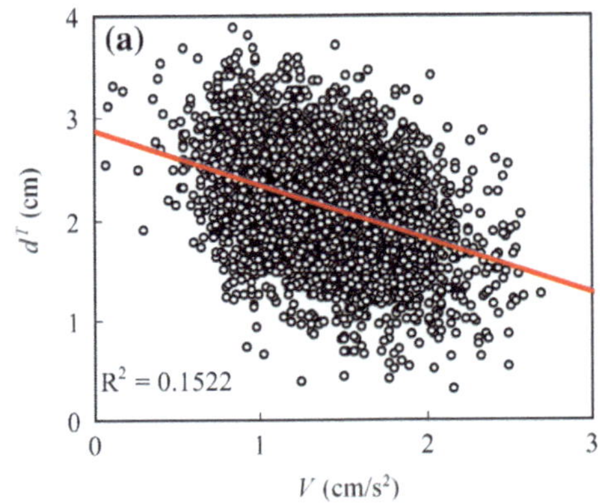

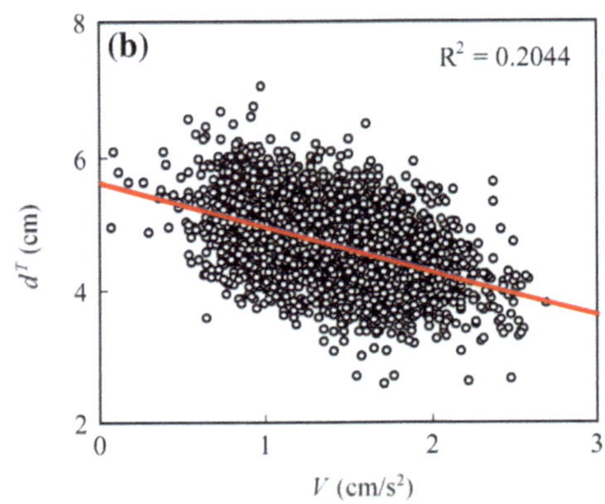

Fig. 3.4 Correlation between displacement and acceleration RMS (hourly averaged): **a** north abutment and **b** south abutment

fic. In this figure, the V-d^T plotting is somewhat dispersed and an overall decrease in displacement can be observed with the increase in acceleration RMS of the bridge. A linear regression analysis between V and d^T was further performed. The expressions of the linear regression function are also shown in Table 3.1.

Before the displacements were utilized to analyze the wind–displacement correlation, the traffic effect on the temperature-normalized displacements d^T should be removed. It was achieved by re-normalizing d^T to a fixed reference acceleration RMS with the use of the presented functions of RMS displacement shown in Table 3.1. The re-normalizing displacements by removing traffic effect were obtained by

$$d^V = d^T - (d_V - d_{VR}) \tag{3.2}$$

where d^V is the re-normalized displacement; d_V is the traffic-induced displacement calculated by feeding the acceleration RMS into the regression functions; d_{VR} is the nominal displacement with the reference RMS value 1 cm/s^2.

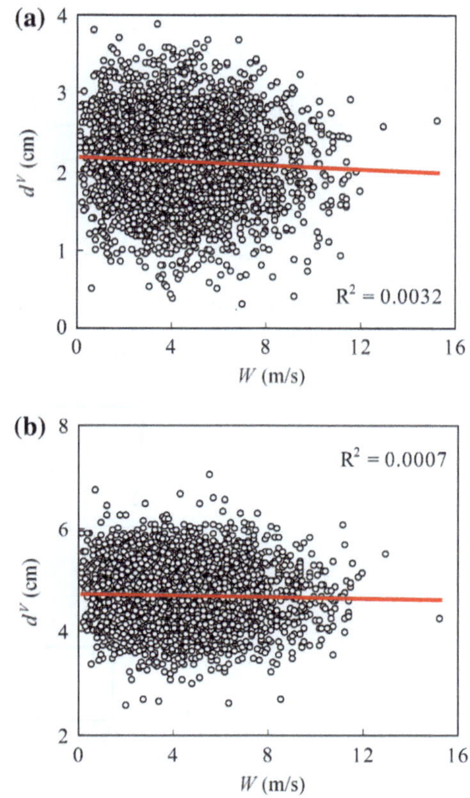

Fig. 3.5 Correlation between displacement and wind speed (hourly averaged): **a** north abutment and **b** south abutment

Figure 3.5 shows the scatter diagrams of re-normalized displacement versus wind speed. In this figure, W represents wind speed and the range of W is (0.15, 15.26 m/s). It can be observed that the W-d^V plotting is rather scattered for both two expansion joints and the correlation between displacement and wind speed is pretty weak. A linear regression analysis was also performed to obtain the correlation models between displacement and wind speed.

In order to get a quantitative analysis of the effects of above three environmental factors on the expansion joint displacement, the range of the environmental factors was fed to the regression functions shown in Table 3.1. The results are listed in Table 3.2.

It can be concluded that (i) Temperature is the critical source causing displacement variability, and there is a linear increase in displacement with the temperature increase; (ii) Although the RMS-displacement correlation is weaker than temperature–displacement correlation, it cannot be ignored. In addition, there is an overall decrease in displacement with the increase in RMS; (iii) The correlation between displacement and wind speed is so weak that the effect of wind on displacement can be

Table 3.2 Variation of displacement induced by environmental factors (unit: cm)

Environmental factors	Temperature		Traffic		Wind	
	North	South	North	South	North	South
Max value	18.95	21.31	2.84	5.57	2.20	4.72
Min value	−19.28	−17.11	1.42	3.81	1.99	4.61
Range (max value–min value)	38.23	38.42	1.42	1.76	0.21	0.11

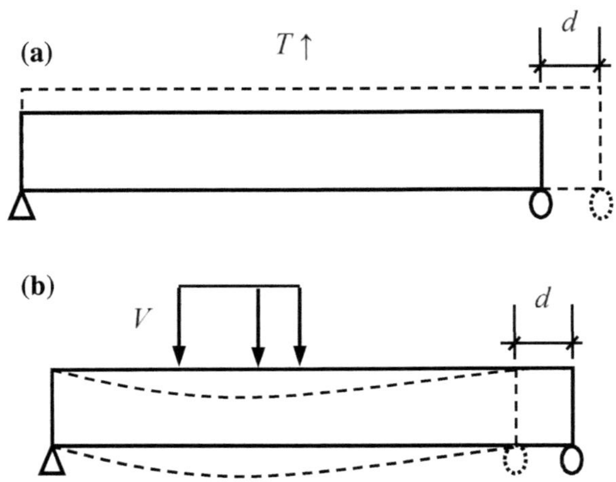

Fig. 3.6 Schematic diagram of deformation of a simply supported beam: **a** temperature rise and **b** traffic loads

ignored. In short, temperature and traffic condition are the dominant environmental factors of the expansion joint displacement.

To present a rational physical interpretation of the influence law of temperature and traffic on the expansion joint displacement, herein a simply supported beam, which had similar support conditions with the girder of a suspension bridge, was employed. Figures 3.6a, b present the deformation of a simply supported beam influenced by temperature increase and traffic loads, respectively. In these figures, the solid line represents the initial state and the dashed line represents the deformation state. Figure 3.6a shows that when the temperature gets higher, the length of the simple supported beam gets longer due to the thermal expansion. Figure 3.6b shows that traffic loading gets bigger, the length of the beam gets shorter due to the increase of the transverse deflection. Figure 3.6 can give a consistent explanation for the temperature–displacement and traffic–displacement relationships shown in Figs. 3.3 and 3.4.

3.4 Damage Detection of Expansion Joints

3.4.1 Correlation Models Between Displacements and Dominant Environmental Factors

According to the previous work, it was clear that there was an inherent pattern of the correlation model between the expansion joint displacements and the dominant environmental factors while the expansion joints were healthy. When expansion joints become damaged due to corrosion and aging, the correlation model may stray from the one in the healthy state. Thus, the first step of the damage detection method was to establish the correlation models between displacements and dominant environmental factors.

The 130-day data, of which the length was 3120, was divided into two parts. Four-fifth of the data was the training part and one-fifth of the data was the testing part. Note that, there was no overlap between the training and testing parts. The data in the training part was used to obtain the multiple linear regression models between the expansion joint displacements and the dominant environmental factors. For north abutment, the regression model is

$$d = -17.30 + 1.01T - 0.21V \tag{3.3}$$

For south abutment, the model is

$$d = -14.72 + 1.02T - 0.60V \tag{3.4}$$

In both equations, d is the hourly expansion joint displacement. T and V denotes temperature and acceleration RMS, respectively.

3.4.2 Definition of Damage Detection Index

First of all, the initial damage detection index e (hourly averaged) was defined as the difference between the measured displacement and the simulated displacement calculated by feeding the measured T and V into the regression models shown as Eqs. (3.3) and (3.4) for north abutment and south abutment, respectively

$$\mathrm{e} = d_M - d_s \tag{3.5}$$

where d_M is the hourly measured displacement and d_S is the simulated displacement.

Taking the north abutment as an example, two kinds of damage detection indices are presented in Fig. 3.7. Figure 3.7a shows the initial damage detection index, which is hourly averaged value. The advanced damage detection index, which was daily averaged from the initial damage detection index, is shown in Fig. 3.7b. So,

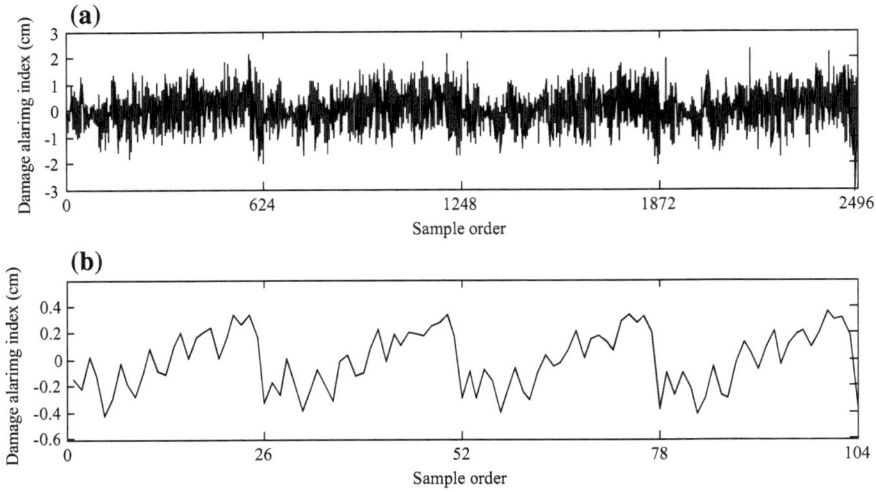

Fig. 3.7 Damage detection indices of the training data from the north abutment: **a** hourly averaged value and **b** daily averaged value

Table 3.3 Statistical characteristics of the damage detection indices of training data (unit: cm)

Index	Hourly averaged index		Daily averaged index	
Location	North abutment	South abutment	North abutment	South abutment
Max value	2.33	2.47	0.36	0.52
Min value	−3.31	−2.54	−0.42	−0.44
Mean value	0.00	0.00	0.00	0.00
Standard deviation	0.65	0.67	0.21	0.25

the lengths of the initial damage detection index and the advanced ones are 2496 and 104, respectively. Furthermore, Table 3.3 gives statistical characteristics of the damage detection indices of training data. It is found that the inherent randomness in the damage detection index is greatly eliminated after the daily averaged calculation, so the advanced index may be more suitable than the initial index in damage detection for RSBs expansion joints.

3.4.3 Statistical Control Chart

Statistical process control is a tool of statistical quality control to detect if the process is out of control. It plots the quality characteristic as a function of the sample number. The chart has lower and upper control limits, which are computed from those samples only when the process is assumed to be in control. When unusual sources of variability are present, sample statistics will plot outside the control limits. In that occasion, an alarm is triggered. There exist different control charts, differing on their plotted statistics. Several univariate and multivariate control charts for damage detection were studied in the past [14–16].

To detect a change in the mean of the damage detection index, an intuitively appealing idea is to form rational subgroups of size n, compute the sample mean within each subgroup and chart the sample means [14].

There are two damage detection indices from north and south abutments, so the size of each subgroup n is 2. And, the lengths of the training data and the testing data are 104 and 26. The number of subgroups m is 104 when only the training data are feeding to X-bar chart. The number of subgroups m will be 130 when both training and testing data are feeding to X-bar chart. The mean value and variance of each subgroup are

$$\overline{X}_i = \frac{\sum_{j=1}^{n} X_{ij}}{n} \quad (i = 1, \ldots, m; j = 1, \ldots, n) \tag{3.6}$$

$$S_i^2 = \frac{\sum_{j=1}^{n} \left(X_{ij} - \overline{X}_i\right)^2}{n} \quad (i = 1, \ldots, m;\ j = 1, \ldots, n) \tag{3.7}$$

where $\overline{X}_i$ is the charted value. The centerline of the chart is the mean of the charted value. The expression of the centerline is

$$\mathrm{CL} = \frac{\sum_{i=1}^{m} \sum_{j=1}^{n} X_{ij}}{mn} \quad (i = 1, \ldots, m; j = 1, \ldots, n) \tag{3.8}$$

The control limits are defined as

$$\mathrm{UCL}, \mathrm{LCL} = \mathrm{CL} \pm Z_{\alpha/2} \frac{S}{\sqrt{n}} \tag{3.9}$$

where UCL and LCL are upper and lower control limits, respectively; α is the confidence level; Z_α represents the α quantile of the standard normal distribution; and the value of S is $\sum_{i=1}^{m} S_i/m$.

3.4.4 False Positive Tests

Although it is desirable to have features sensitive to damage, the monitoring system also needs to be robust against a false positive indication of damage; false positive indication of damage means that the monitoring system indicates damage although no damage is present. To investigate the robustness of control chart, two separate tests were designed with different α. In both tests, the expansion joints were healthy. The false positive test results are shown in Figs. 3.8 and 3.9. The subgroup sample inside the control limits is marked by a “o”. Meanwhile, outliers correspond to subgroups

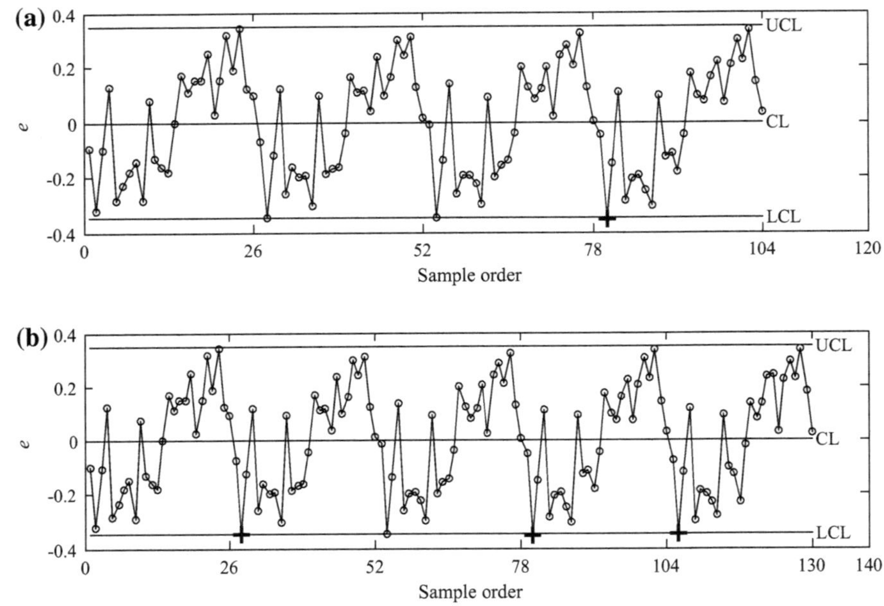

Fig. 3.8 Control charts with false positive indications of damage: **a** only training data and **b** both training data and test data

sample means outside the control limits and are marked by a "+". Figure 3.8 shows the control charts with false positive indications of damage. Because α is 0.004 in Fig. 3.8, approximately no charted values (0.004 of total 104 or 130 samples) are expected to be outside the control limits. In other words, if any one of the subgroup samples exceeds the control limits it means a false positive indication of damage. As shown in Fig. 3.8, there is an outlier in the chart established with only training data and there are 3 outliers in the chart established with both training and testing data. The outliers in this figure show the classic false positive indications of damage. And the damage detection will be wrongly triggered inevitably.

Then, the value of α was changed from 0.004 to 0.003 and the new control charts are shown in Fig. 3.9. In both Fig. 3.10a, b, all the subgroup samples are inside the control limits. It means that there is no false positive indication of damage because of the proper choice of α. The false positive tests provide evidence that the X-bar control chart is robust against false positive indication of damage and can effectively reduce the probability of misjudgment by the multi-sample hypothesis test. The value of α continues to be 0.003.

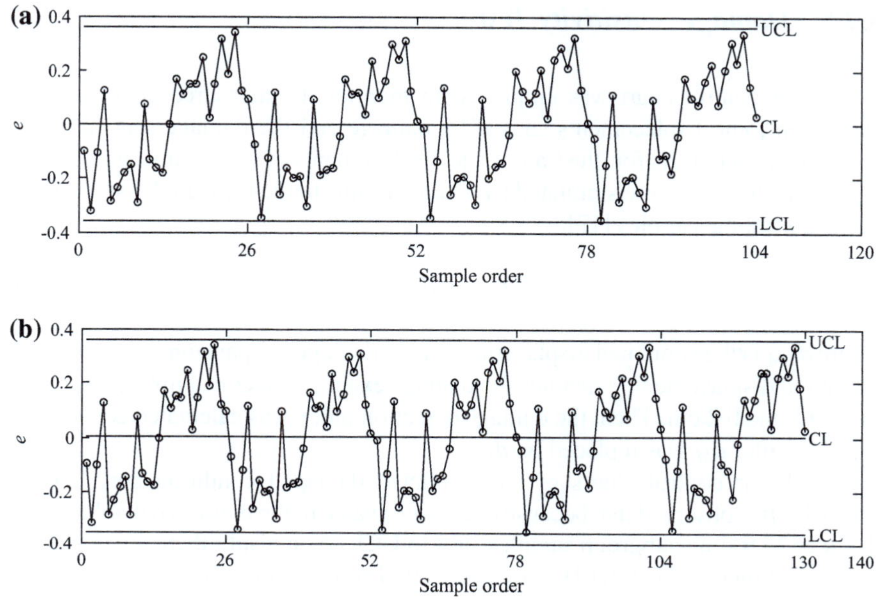

Fig. 3.9 Control charts without false positive indications of damage: **a** only training data and **b** both training data and test data

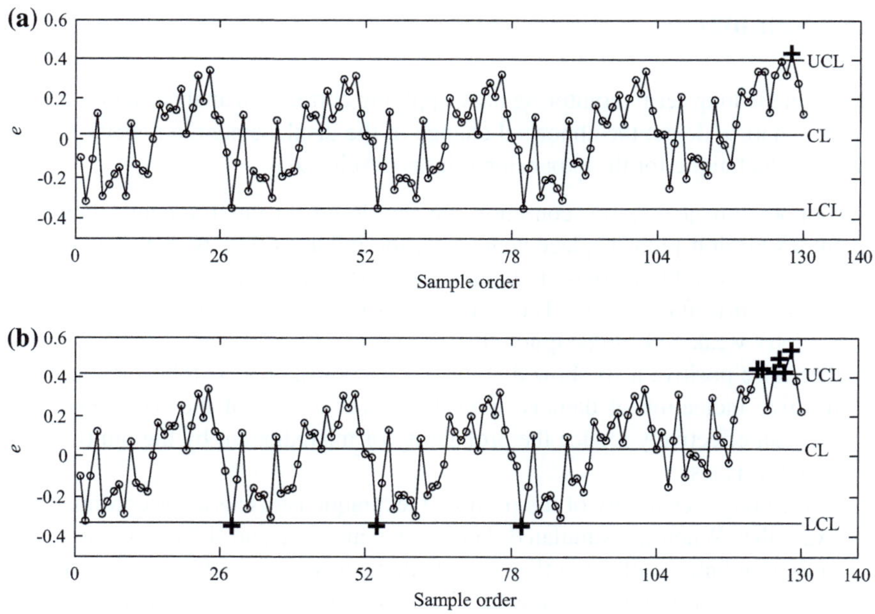

Fig. 3.10 Control charts of damage expansion joints: **a** $\varepsilon=0.1$ cm and **b** $\varepsilon=0.2$ cm

3.4.5 Damage Sensitivity Test

The X-bar control chart was employed to monitor the measured changes of the expansion joint displacements caused by damage and the damage sensitivity was discussed in detail. As for illustration, the displacements of the expansion joint under damaged condition were simulated by adding a value to the measured displacements of the 26 testing samples [17]:

$$d_{\mathrm{DM}} = d_M + \varepsilon \tag{3.10}$$

where d_{DM} is the simulated displacement of the damaged expansion joints; d_M is the measured displacement; ε denotes the damage extent, in later examples, ε is chosen to be 0.1 and 0.2 cm. Then, the damage detection index was calculated according to Eq. (3.5) after d_M was replaced by d_{DM}.

The X-bar control charts of the simulated damaged condition are shown in Fig. 3.10. It is observed that because of adding a value to the measured displacements the later 26 testing subgroup samples shift to UCL. As a result, there are one outlier and nine outliers in Figs. 3.10a, b, respectively. These outliers will trigger the damage detection of the expansion joints.

3.5 Summary

Based on the long-term monitoring data, applying correlation analysis and statistical control chart to detect the abnormal change of the displacement, a novelty damage detection technique for the expansion joint is developed:

1. Temperature and traffic condition are the dominant environmental factors of the expansion joint displacement. And, the influence law is that displacement increases with the increase in temperature and displacement decreases with the increase in traffic intensity. The correlation between displacement and wind speed is pretty weak and can be ignored.
2. The false positive tests show that the X-bar control chart is robust against false positive indication of damage with the proper choice of the confidence level and can effectively reduce the probability of misjudgment by the multi-sample hypothesis test.
3. The damage sensitivity of the proposed technique is discussed according to two examples. When the simulated damage extent is 0.1 and 0.2 cm, there are one outlier and nine outliers in the control charts, respectively. Hence, the proposed technique can effectively detect the slight abnormal change of the expansion joint displacements and it is suitable for the real-time structural health monitoring of the RSBs expansion joints.

References

1. Ko JM, Ni YQ. Technology developments in structural health monitoring of large-scale bridges. Eng Struct. 2005;27(12):1715–25.
2. Doebling SW, Farrar CR, Prime MB. A summary review of vibration-based damage identification methods. Shock Vib Dig. 1998;30(2):91–105.
3. Deng Y, Ding YL, Li AQ. Fatigue reliability assessment for bridge welded details using long-term monitoring data. Sci China Technol Sci. 2011;54(12):3371–81.
4. Deng Y, Ding YL, Li AQ. Prediction of extreme wind velocity at the site of the Runyang Suspension Bridge. J Zhejiang Univ Sci A. 2011;12(8):605–15.
5. Ni YQ, Hua XG, Fan KQ, Ko JM. Correlating modal properties with temperature using long-term monitoring data and support vector machine technique. Eng Struct. 2005;2(12):1762–73.
6. Ding YL, Deng Y, Li AQ. Study on correlations of modal frequencies and environmental factors for a suspension bridge based on improved neural networks. Sci China Technol Sci. 2010;53(9):2501–9.
7. Cornwell P, Farrar CR, Doebling SW, Sohn H. Environmental variability of modal properties. Exp Tech. 1999;23(6):45–8.
8. Ding YL, Li AQ. Temperature-induced variations of measured modal frequencies of steel box girder for a long-span suspension bridge. Int J Steel Struct. 2011;11(2):145–55.
9. Chen WF, Duan L. Bridge engineering handbook. Boca Raton: CRC Press; 2000.
10. Ni YQ, Hua XG, Wong KY, Ko JM. Assessment of bridge expansion joints using long-term displacement and temperature measurement. J Perform Constr Facil. 2007;21(2):143–51.
11. Li AQ, Miao CQ, Li ZX, Han XL, Wu SD, Ji L, Yang YD. Health monitoring system for the Runyang Yangtse River Bridge. J Southeast Univ (Nat Sci Edit). 2003;23(5):544–8.
12. Zhang QW, Fan LC, Yuan WC. Traffic-induced variability in dynamic properties of cable-stayed bridge. Earthq Eng Struct Dynam. 2002;31(11):2015–21.
13. Chen J, Xu YL, Zhang RC. Modal parameters identification of Tsing Ma suspension bridge under typhoon Victor: EMD-HT method. J Wind Eng Ind Aerodyn. 2004;92(10):805–27.
14. Fugate LM, Sohn H, Farrar CR. Vibration-based damage detection using statistical process control. Mech Syst Signal Process. 2001;15(4):707–21.
15. Kullaa J. Damage detection of the Z24 bridge using control charts. Mech Syst Signal Process. 2003;17(1):163–70.
16. Deraemaeker A, Reynders E, De Roeck G, Kullaa J. Vibration-based structural health monitoring using output-only measurements under changing environment. Mech Syst Signal Process. 2008;22(1):34–56.
17. Miao CQ, Deng Y, Ding YL, Li AQ. Damage alarming for bridge expansion joints using novelty detection technique based on long-term monitoring data. J Cent South Univ. 2013;20(1):226–35.

Chapter 4
Modal Frequency-Based Structural Damage Detection

4.1 Introduction

Over the past several decades, a significant research effort has focused on the health monitoring and condition assessment for long-span bridges [1, 2]. How to explain the health condition of the bridge structure according to the collected structural responses remains a great challenge in the civil engineering community. It is well known that bridge structures are subject to varying environmental conditions such as traffic loadings and environmental temperature. These environmental effects will cause changes in the structural damage detection parameters which may mask the changes caused by structural damage. Therefore, for reliable performance of structural health monitoring technique for long-span bridges, it is of paramount importance to characterize normal variability of damage detection parameters due to environmental effects and discriminate such normal variability from abnormal changes in damage detection parameters caused by structural damage [3, 4].

Considerable research efforts have been devoted to investigating the influences of environmental conditions on modal frequencies of bridges. Most of these investigations have indicated that temperature is the most significant environmental effect affecting bridge modal properties [1, 3, 5–9]. For instance, Cornwell et al. [5] observed the variability of modal frequencies by up to 6% over a 24-h period on the Alamosa Canyon Bridge. Wahab and De Roeck [8] conducted two dynamic tests for a prestressed concrete bridge in spring and winter, respectively, and observed an increase of 4–5% in modal frequencies with the decrease in temperature. In addition, some reported studies have shown that vehicle mass has very little influence on the modal parameters of large-scale bridges [10, 11]. And the modal frequencies had a slight decrease with the increase of wind speed [12, 13].

It should be noted that most of these studies observed the variations in measured modal frequencies under normal environmental conditions; however, the modal variability induced by temperature, traffic loadings, and wind speed were not simultaneously quantitatively evaluated. And, it is more important that all the studies did not

Y. Deng and A. Li, *Structural Health Monitoring for Suspension Bridges*,
https://doi.org/10.1007/978-981-13-3347-7_4

take into consideration the measurement errors arising from the modal identification algorithm. For instance, Ni et al. [13] observed that the normal environmental change of Ting Kau Bridge accounts for relative variation in modal frequencies from 3.22 to 15.07% for the first ten modes. Using the formulated frequency–temperature correlation models, the relative variation of the normalized frequencies after removing the temperature effect was reduced to range from 1.61 to 7.87%. In their studies, the correlation between frequency and wind speed is very weak and the wind effect on the modal frequencies is of no significance. Thus, one possible reason for the considerable variations in the normalized frequencies after removing the temperature and wind effect is that the measurement errors induced by the modal identification algorithm are significant. However, relevant works are seldom reported.

After the environmental or temperature effects are eliminated or reduced by using the correlation models, the baseline database could be established based on the modal frequencies from the healthy phase of the bridge structures. However, there is randomness in the modal frequencies even after elimination of the environmental effects [14, 15]. The probabilistic tool is required to determine structural damage status when comparing the measured data to the baseline database. Hence, many studies adopted the control charts, a tool of statistical quality control to detect if the process is out of control, to detect damage of bridge structures. Various control charts, such as the X-bar [16–19], Shewhart T [20–22], Hotelling T [20, 23], CUSUM [20, 24], and EWMA [19, 20] control charts, have been, respectively, studied in the structural damage detection. It should be mentioned that the control chart assumes that the quality characteristic follows the normal distribution [25]. However, the studies on checking whether the extracted features are normally distributed before using the control charts for damage detection are rare. Quesenberry [26–29] proposed a class of statistics, called Q statistics, to construct Shewhart-type control charts for individual observations from normal, binomial, and Poisson process distributions. In particular, the methodology in reference [27] can convert non-normal observed data to the Q statistic to obey the standard normal distribution using probability integral transformation theorem.

Based on the aforementioned motivation, the first objective of this chapter is to quantify the modal variability of the RSB induced by temperature, wind speed, traffic loadings, and identification algorithm making use of long-term monitoring data. First, the output-only modal frequency identification of the bridge is effectively carried out using the iterative windowed curve-fitting method (IWCM) in the frequency domain. The seasonal temperature–frequency correlation is formulated using multi-sample averaging technique for each vibration mode to eliminate the random variations rising from the identification algorithm. Then, the effect of temperature on the measured modal frequencies is eliminated by normalizing these modal frequencies to a reference temperature with the use of the temperature–frequency correlation models. The traffic-induced modal variability is further quantitatively evaluated under weak wind conditions by correlation the normalized modal frequencies with the root mean square (RMS) of deck vertical acceleration responses. Based on the traffic–frequency correlation models, the effect of traffic loadings on the modal frequencies is eliminated and the wind-induced modal variability is quantitatively evaluated under weak and strong wind conditions.

In addition, a framework based on the control chart and long-term monitoring data is presented for damage detection of long-span bridges using modal frequencies. The backpropagation neural network (BPNN) is adopted to establish the correlation model between the modal frequencies and structural temperature. Then, the kernel density estimate is used to obtain the distribution functions of the modal frequencies after the temperature effect elimination and then the modal frequencies will be transferred into the Q statistics. Finally, the X-bar control chart is applied to detect the abnormality of modal frequencies, which are used as the indices of damage detection in this study. The efficiency of the proposed framework is verified on a prototype bridge. 215-day monitoring data is processed and analyzed to show its advantages.

4.2 Identification of Modal Frequencies for the RSB

The prototype bridge of this chapter is the Runyang Suspension Bridge (RSB). A total of 27 uniaxial servo type accelerometers have been installed at the nine sections of the bridge deck to measure dynamic responses. The nine sections are equidistantly located in the main span of the RSB as shown in Fig. 4.1. Likewise, a total of 27 temperature sensors have been installed at 4 sections (Sections 1, 2, and 9 in Fig. 4.1) of the bridge deck to measure the temperature of steel box girder. For wind environmental monitoring, two WA15 anemometers produced by the Vaisala Company were installed in the SHMS. One is located at the deck level and the other at tower top (Fig. 3.1). The sampling rates for accelerometers, anemometers and temperature sensors are 20, 1 and 1 Hz, respectively.

To provide an accurate and reliable evaluation of the real dynamic characteristics of the RSB, long-term field ambient testing was carried out from January to October in the year 2006 by using the SHMS. During this period, sufficient variation in temperatures, wind velocities and traffic loadings were experienced for any trends in modal parameters to be identified. The peak picking (PP) method in the frequency domain was used for the identification of modal frequencies of the RSB. In the PP method, the identified natural frequencies are simply obtained from the observation of the peaks on the graphs of the auto power spectral densities (PSDs) of the output-only data [30, 31]. In order to overcome the shortcomings of the PP method such as bias errors and limited frequency resolutions, the iterative windowed curve-fitting method

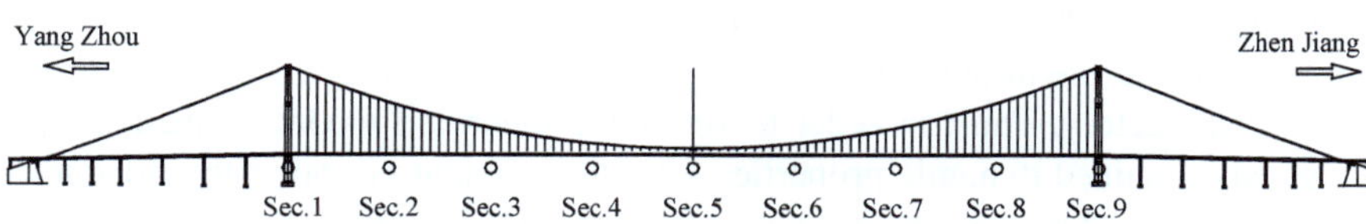

Fig. 4.1 Layout of ambient vibration monitoring of the RSB

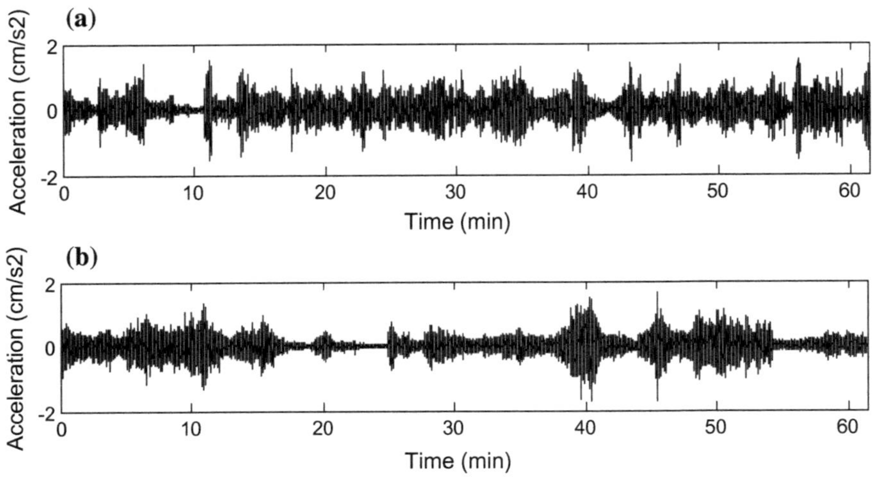

Fig. 4.2 Vertical acceleration time histories of the deck: **a** Section 2 and **b** Section 5

(IWCM) is further used to improve the modal frequency estimates [32]. IWCM is a frequency domain method, based on the curve fitting of PSD, but modified to deal with spectral bias errors. Bias errors are a significant problem in conventional frequency domain methods, causing attenuation of values around spectral peaks in the PSD. This arises from the finite length of data segments used for the fast Fourier transform (FFT) estimation of the PSD. The key to IWCM is that it modifies the modal peaks of the theoretical PSD, to account for the bias, before it is fitted to the measured PSD. This is done by applying the same window function used on the measured data, in the time domain, to the theoretical spectral peaks, in the frequency domain, thus improving the modal frequency estimates. Per 10-min recorded data are used for PSD analysis. Before the PSD analysis is performed, the measured 10-min data are first detrended, which enables the removal of the DC components that may badly influence the identification results. This was accomplished by subtracting the mean values calculated over the 10-min duration of each measurement [30].

Figure 4.2 shows the vertical acceleration time histories of the deck at Sections 5 and 2. It can be observed that the measured dynamic responses present the nonstationary characteristics due to the wind and traffic loadings.

Figure 4.3 shows the measured normalized PSDs of deck acceleration responses and IWCM curve fits to vertical deck PSDs. The good fits suggest that the identified modes describe the dynamic behavior of the structure well. As shown in the figure, the peaks of the PSD plots clearly indicate the modal frequencies of the bridge. However, due to the limited measured sections, the analysis results of mode shapes are not satisfactory. Thus, in order to verify the identified modal frequencies, the previously obtained dynamic properties from the ambient vibration test conducted just prior to the opening of the bridge are further used [33], because in that vibration test, 47 sections are installed accelerometers to obtain both modal frequencies and

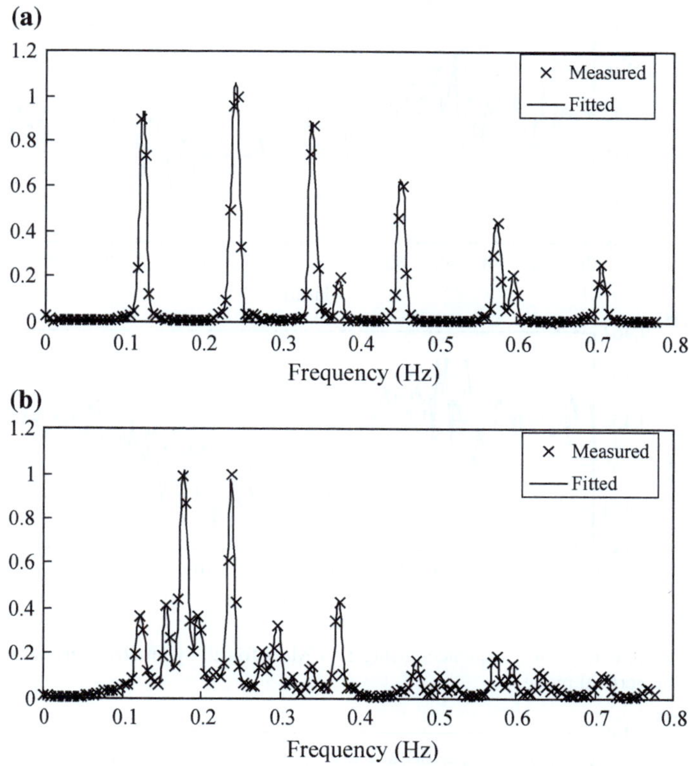

Fig. 4.3 Typical curve-fit results of normalized PSD of vertical deck acceleration: **a** Section 5 and **b** Section 2

mode shapes, which are the basis for the proper determination of modal frequencies from the measured PSD plots such as Fig. 4.3.

Figure 4.4 shows the identified frequency sequences in a typical day for the RSB using the IWCM. On the whole, the measured modal frequencies have the minimum approximately at 2:00 PM in the afternoon and reached to the maximum approximately at 6:00 AM in the wee hours. Therefore, the measured modal frequencies can effectively reflect the fluctuation characteristics of ambient temperature in one day.

As for the comparisons, Fig. 4.5 gives the identification results using the conventional PP method. It can be seen that due to the limited frequency resolutions the identified modal frequency sequences without using IWCM fails to reflect the influence of daily temperature variations. Thus, the identified modal frequencies using IWCM provide an effective indication for changes of the bridge due to the ambient temperature variations. It should be noted that before using IWCM, it is necessary to pick up the peaks on the graphs of the PSDs for the frequency identification. For identified peaks using PP method, the usage of IWCM is to improve the modal frequency identification. It is usually observed that the PP method in the frequency domain cannot identify all important mode shapes for such a large bridge [30]. Thus,

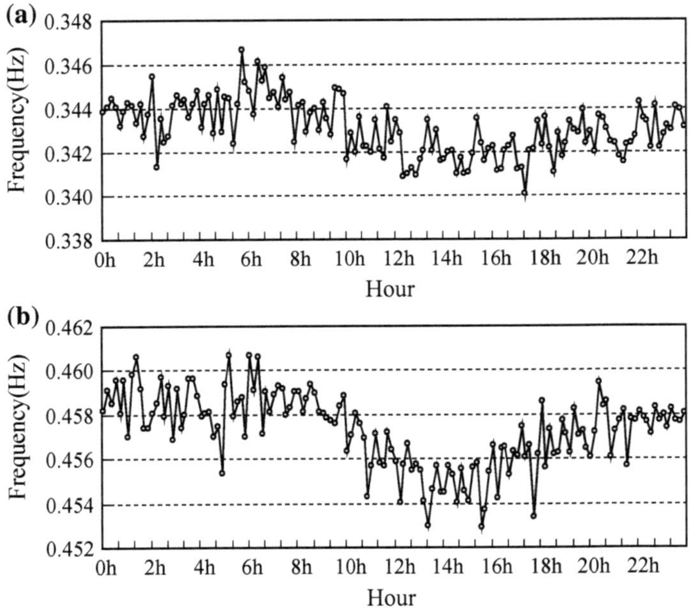

Fig. 4.4 Measured frequency sequences using IWCM: **a** fourth symmetric vertical mode and **b** fifth symmetric vertical mode

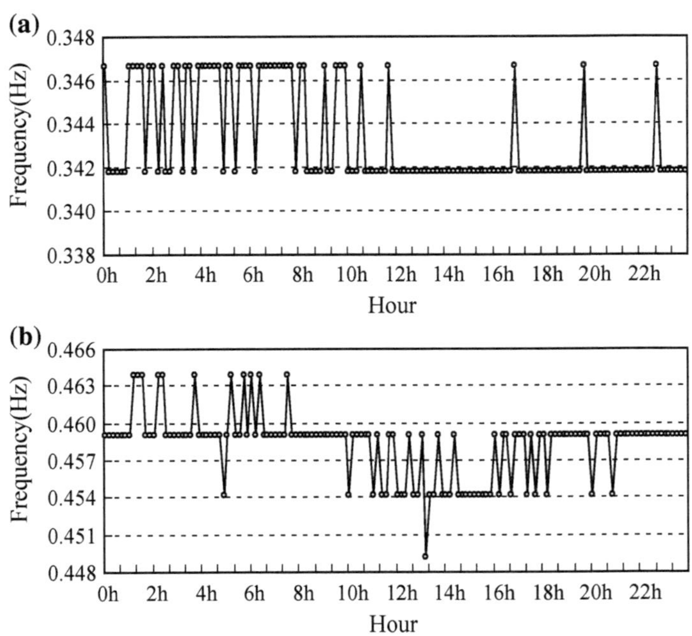

Fig. 4.5 Measured frequency sequences using conventional PP method: **a** fourth symmetric vertical mode and **b** fifth symmetric vertical mode

the IWCM can only be used to overcome the shortcoming of PP method with limited frequency resolutions and cannot be used to obtain complete modal information. In addition, in Fig. 4.4, it can be observed that the influences of ambient loadings on the measured frequencies present instantaneous changes because the modal identification algorithm in the frequency domain assumes stationarity of loading and hence response. However, the measured dynamic responses of the bridge were rejected for statistically meaningful stationarity tests. Hence, the modal frequency variation induced by the modal identification algorithm is mainly random variations due to nonstationary ambient loadings.

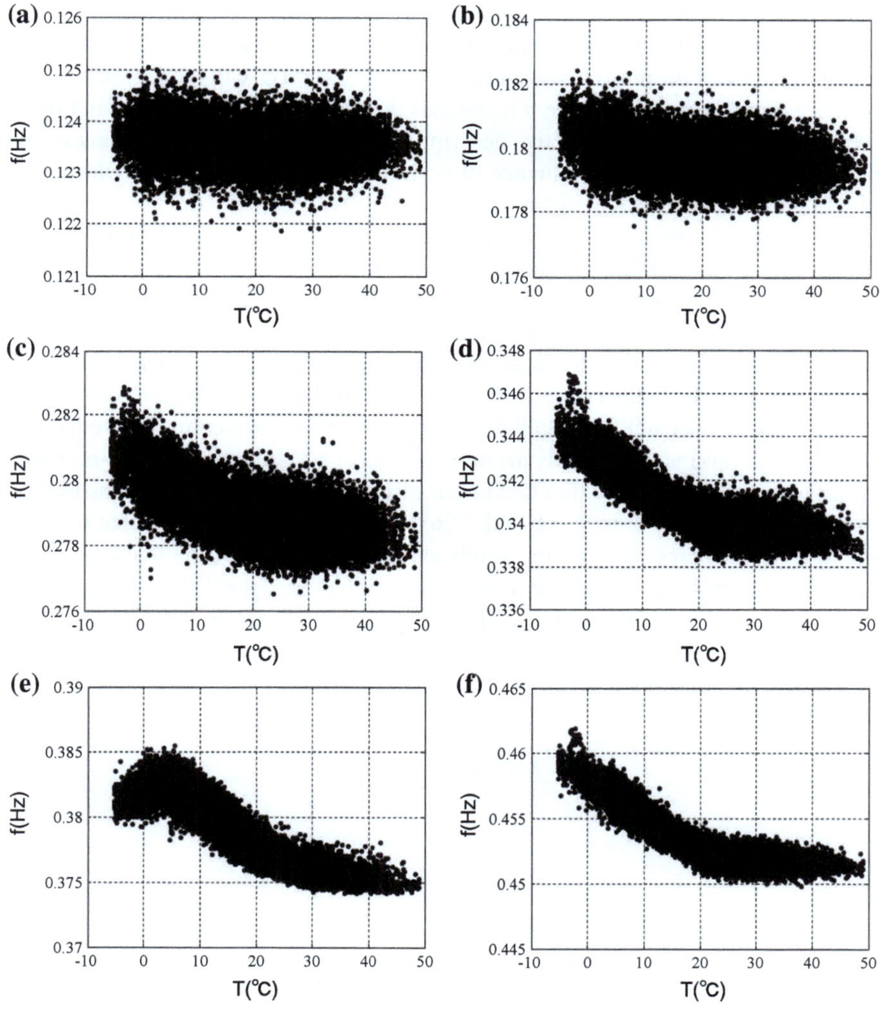

Fig. 4.6 Correlations of frequency–temperature: **a** Mode 1, **b** Mode 2, **c** Mode 3, **d** Mode 4, **e** Mode 5, and **f** Mode 6

4.3 Temperature-Induced Variability of Modal Frequencies

4.3.1 Correlation Analysis of Temperature–Frequency

The 215-day acceleration measurement data (from January to October in the year 2006), after removing those likely to be abnormal are used in this study. With the acceleration measurement data, the modal frequencies of the six vibration modes were identified at 10-min intervals. Hence, there were a total of 30,960 (144×215) samples of modal frequency for each vibration mode. Figure 4.6 plots modal frequencies versus temperature with regard to different vibration modes, in which the temperature data from all sensors were averaged to a value as the representative temperature of the bridge. All six vibration modes are listed in Table 4.1. For all the six modes, an overall decrease in modal frequency is observed with the increase in temperature of the bridge. However, it is obvious that the measurement points in Fig. 4.6 are too dispersed to be an effective description of the correlation of frequency–temperature, implying that the influence of nonstationary loading on the identification of modal frequencies is significant.

The seasonal correlation analysis is further applied to eliminate the random variations due to nonstationary loadings, namely using the daily averaged values to construct the seasonal relationship between frequency and temperature. In Fig. 4.7, the daily averaged frequency–temperature scatter diagrams are presented. It can be shown that measured modal frequencies of vibration modes have remarkable seasonal correlation with the temperature. Table 4.1 summarizes the statistical information of modal frequencies from 215-day data in the year 2006, in which modal frequencies at 1-day intervals means the daily averaged values of 144 frequencies at 10-min intervals. It can be observed that the maximum and averaged relative variations reduced from 3.013 and 2.633% to 2.168 and 1.268%, respectively, implying that the daily averaged frequencies using multi-sample averaging technique can effectively eliminate the random variations rising from the identification algorithm.

A 6-order polynomial regression model is further applied herein for the modeling of frequency–temperature seasonal correlations. The seasonal correlations can be mathematically described as

$$f(T) = \sum_{i=0}^{6} p_i \times T^i \tag{4.1}$$

Table 4.1 Statistics of measured modal frequencies at 10-min and 1-day intervals

Mode No.	Mode description	10-min interval				1-day interval			
		Maximum (Hz)	Minimum (Hz)	Average (Hz)	Relative variation (%)	Maximum (Hz)	Minimum (Hz)	Average (Hz)	Relative variation (%)
1	First symmetric vertical mode	0.1250	0.1218	0.1235	2.591	0.1238	0.1233	0.1235	0.405
2	Second anti-symmetric vertical mode	0.1824	0.1776	0.1797	2.671	0.1807	0.1792	0.1797	0.835
3	Third anti-symmetric vertical mode	0.2829	0.2765	0.2790	2.294	0.2809	0.2780	0.2790	1.039
4	Fourth symmetric vertical mode	0.3469	0.3381	0.3408	2.582	0.3441	0.3392	0.3408	1.438
5	Fourth anti-symmetric vertical mode	0.3855	0.3741	0.3783	3.013	0.3833	0.3751	0.3783	2.168
6	Fifth symmetric vertical mode	0.4618	0.4498	0.4535	2.646	0.4588	0.4510	0.4535	1.720

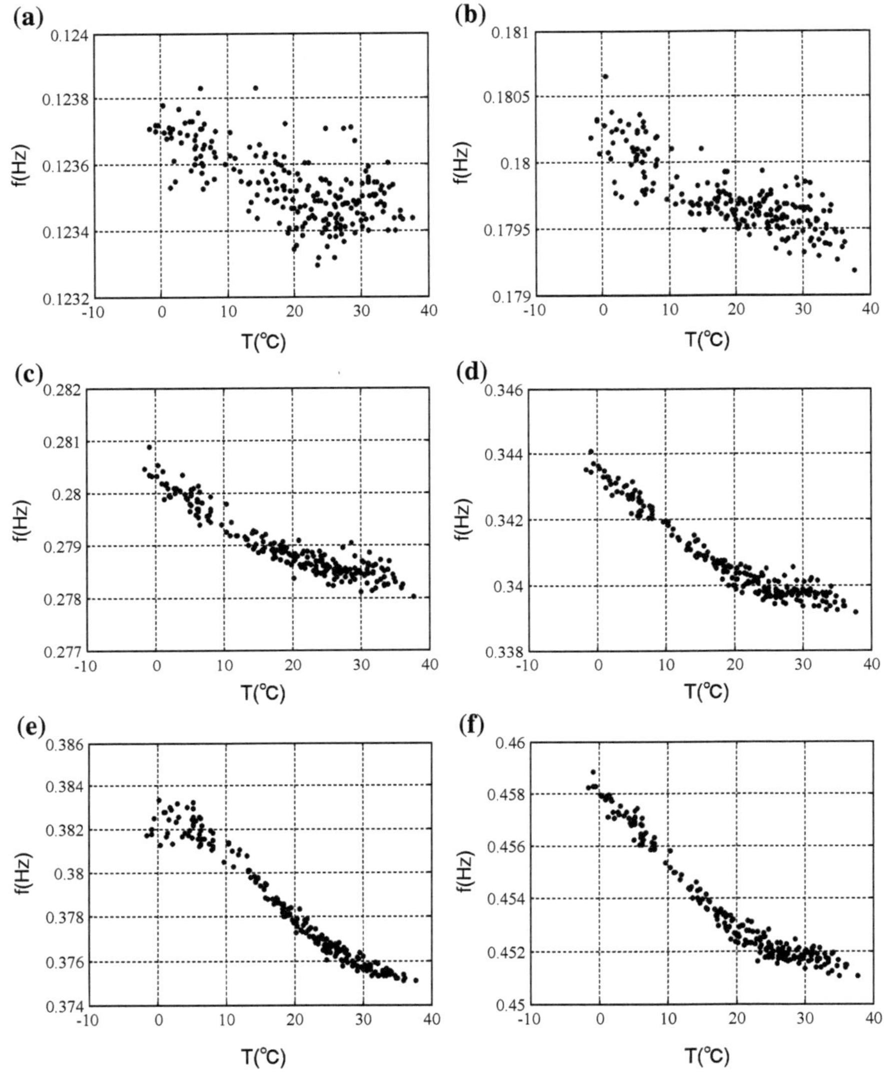

Fig. 4.7 Seasonal correlations of frequency–temperature: **a** Mode 1, **b** Mode 2, **c** Mode 3, **d** Mode 4, **e** Mode 5, and **f** Mode 6

Table 4.2 Summary of polynomial regression models

Mode	p_0	p_1	p_2	p_3	p_4	p_5	p_6
1	0.1237	-7.9556×10^{-6}	2.8113×10^{-6}	-3.8560×10^{-7}	1.6987×10^{-8}	-2.7442×10^{-10}	1.1285×10^{-12}
2	0.1802	3.4614×10^{-6}	-2.3921×10^{-6}	-2.3168×10^{-7}	2.5466×10^{-8}	-7.3456×10^{-10}	6.6960×10^{-12}
3	0.2804	-7.4533×10^{-5}	1.4373×10^{-6}	-4.3476×10^{-7}	2.7526×10^{-8}	-5.8201×10^{-10}	3.6489×10^{-12}
4	0.3435	-1.3557×10^{-4}	3.6787×10^{-6}	-1.0614×10^{-6}	6.3712×10^{-8}	-1.3241×10^{-9}	8.5268×10^{-12}
5	0.3822	1.8027×10^{-4}	-4.1850×10^{-5}	1.4321×10^{-6}	-1.8955×10^{-8}	1.5204×10^{-10}	-1.5680×10^{-12}
6	0.4581	-2.0162×10^{-4}	5.5445×10^{-6}	-2.0896×10^{-6}	1.3731×10^{-7}	-3.2575×10^{-9}	2.6275×10^{-11}

where T is the daily averaged value of temperature; f is the daily averaged values of the frequency; $p_0(i = 1\text{–}6)$ is the coefficient of the regression model. The 215-day monitoring data are used for training the polynomial model. Table 4.2 summarizes the parameters of polynomial regression functions.

Figure 4.8 shows the comparisons of the measured and polynomial reproduction results using 215-day training data. In the figure, solid line represents the measured values of modal frequency and dashed line represents the reproduction values using the polynomial regression model. The reproduction values of modal frequency favorably agree with the measured values, which indicate the satisfactory reproduction capability of the regression model. On the whole, the developed polynomial regression model exhibits good capabilities for mapping between the temperature and measured modal frequency so that the temperature-caused variability of the modal frequency can be effectively quantified. Based on the polynomial regression functions, it is observed that the maximum and averaged relative variations of frequency induced by temperature are 1.975 and 1.175% for the variation of temperature covering a full cycle of in-service conditions, as illustrated in Table 4.3.

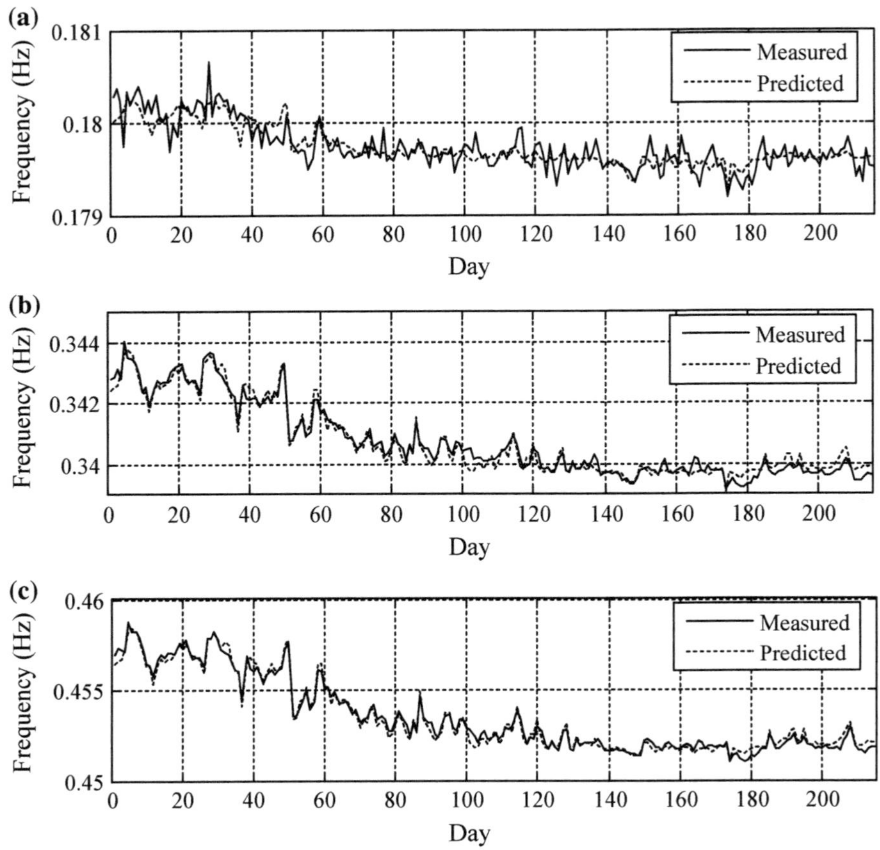

Fig. 4.8 Comparisons of the measured and polynomial reproduction results of frequencies: **a** Mode 2, **b** Mode 4, and **c** Mode 6

Table 4.3 Variations of frequency induced by temperature

Mode No.	Frequency at 0 °C (Hz)	Frequency at 40 °C (Hz)	Relative variation (%)
1	0.1237	0.1232	0.397
2	0.1802	0.1791	0.621
3	0.2804	0.2777	0.965
4	0.3435	0.3385	1.474
5	0.3822	0.3747	1.975
6	0.4581	0.4507	1.620

4.3.2 *Removal of Temperature Effect*

Before the measured frequencies are used for structural health monitoring, the temperature effect on the measured modal frequencies should be removed. It is achieved by normalizing all the measured frequencies to a fixed reference temperature with the use of the established seasonal correlation models for all 6 modes. In this study, the reference temperature is taken as 20 °C. By presenting the reference temperature into the correlation models, a nominal frequency (f_r) is obtained for each vibration mode. Likewise, by feeding the temperature measurement data into the model, a temperature-induced frequency (f_t) is predicted. Then, the normalized frequency after removing temperature effect can be obtained by

$$f = f_m - (f_t - f_r) \tag{4.2}$$

where f is the normalized modal frequency; f_m is the measured modal frequency. It should be noted that modal frequencies measured at both 10-min and 1-day intervals are normalized using the temperature–frequency seasonal correlation models. Figures 4.9 and 4.10 show the measured and normalized modal frequencies at 10-min and 1-day intervals for the second, fourth, and sixth modes, respectively.

Table 4.4 provides the statistics of the normalized modal frequencies at 10-min and 1-day intervals. By comparing the relative variations shown in Table 4.4 with those given in Table 4.1, it is concluded that: (i) The temperature effect on the daily averaged frequencies at 1-day intervals can be effectively removed. The maximum and averaged relative variations reduced from 2.168 and 1.268% to 0.582 and 0.413%. (ii) The elimination of temperature effect on the measured frequencies at 10-min intervals is not satisfactory. The maximum and averaged relative variations reduced from 3.013 and 2.633% to 2.506 and 2.027%, implying that the influence of nonstationary loading on the identification of modal frequencies is significant. Figure 4.9 illustrates the random variations of the normalized modal frequencies at 10-min intervals. Thus, seasonal correlation models of frequency–temperature can effectively eliminate the temperature effect and random variations rising from the identification algorithm and it is suitable for online structural health monitoring and damage warning.

4.4 Wind-Induced and Traffic-Induced Modal Variability

4.4.1 *Correlation Analysis of Traffic–Frequency*

Correlation analysis is conducted on the normalized modal frequencies and traffic conditions. The root mean square (RMS) of the vertical acceleration responses is calculated in the frequency band of 0–3 Hz and RMS data obtained under weak wind conditions (10-min averaged wind speed less than 2 m/s) are mainly due to

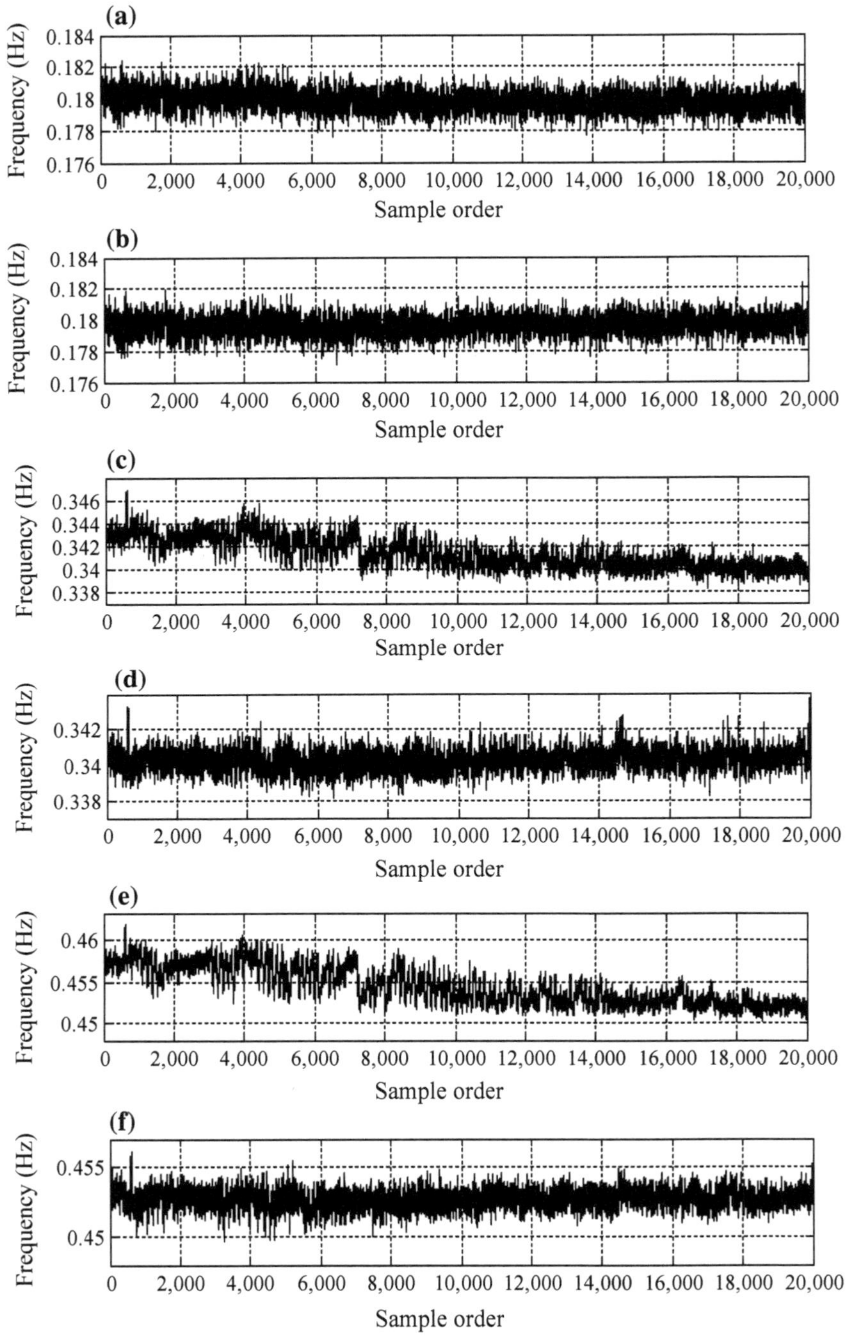

Fig. 4.9 Measured and normalized modal frequencies at 10-min intervals (in part): **a** measured frequencies of mode 2, **b** normalized frequencies of mode 2, **c** measured frequencies of mode 4, **d** normalized frequencies of mode 4, **e** measured frequencies of mode 6, and **f** normalized frequencies of mode 6

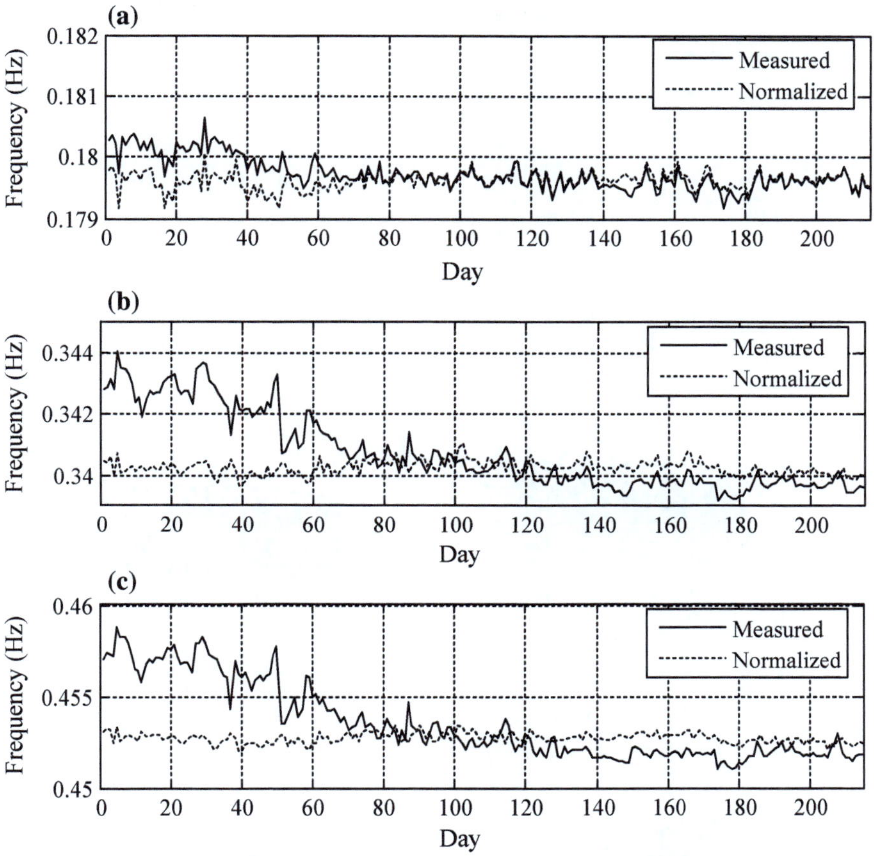

Fig. 4.10 Measured and normalized modal frequencies at 1-day intervals: **a** Mode 2, **b** Mode 4, and **c** Mode 6

the varying traffic loadings and are utilized to characterize the traffic–frequency correlation. The normalized modal frequencies of all 6 modes at 10-min intervals and 10-min acceleration RMS values measured at the deck level (Section 5 in Fig. 4.1) are used. Figure 4.11 shows the correlation diagrams of normalized frequency versus acceleration RMS for the first, third, fourth, and sixth modes, respectively. It can be observed that the RMS frequency plotting is rather scattered and no apparent pattern exists.

A linear regression analysis between the normalized frequency, f, and the RMS value, M, is performed by assuming [12, 13]

$$f = \beta_0 + \beta_1 \times M \tag{4.3}$$

where the regression coefficients β_0 and β_1 are obtained by the least squares method as

Table 4.4 Statistics of normalized modal frequencies at 10-min and 1-day intervals

Mode No.	10-min intervals				1-day intervals			
	Maximum (Hz)	Minimum (Hz)	Average (Hz)	Relative variation (%)	Maximum (Hz)	Minimum (Hz)	Average (Hz)	Relative variation (%)
1	0.1249	0.1219	0.1235	2.429	0.1237	0.1233	0.1235	0.324
2	0.1816	0.1771	0.1796	2.506	0.1801	0.1792	0.1796	0.501
3	0.2815	0.2755	0.2787	2.153	0.2793	0.2783	0.2787	0.359
4	0.3437	0.3381	0.3403	1.646	0.3410	0.3397	0.3403	0.382
5	0.3814	0.3738	0.3779	2.011	0.3790	0.3768	0.3779	0.582
6	0.4561	0.4497	0.4527	1.414	0.4535	0.4520	0.4527	0.331

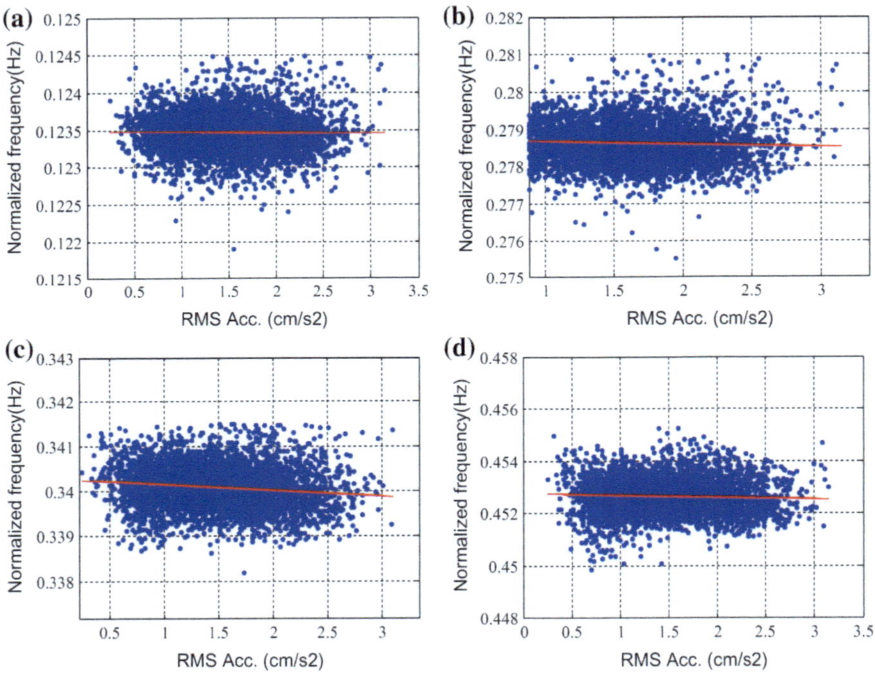

Fig. 4.11 Correlation between normalized modal frequency and RMS of the acceleration responses: **a** Mode 1, **b** Mode 3, **c** Mode 4, and **d** Mode 6

$$\beta_1 = \frac{S_{fM}}{S_{MM}} \tag{4.4}$$

$$\beta_0 = \bar{f} - \beta_1 \overline{M} \tag{4.5}$$

where S_{fM} is the covariance between the normalized frequency and RMS sequences; S_{MM} is the variance of the measured RMS sequences; $\overline{M}$ and $\bar{f}$ are the means of the measured RMS and normalized frequency sequences, respectively. In addition,

Table 4.5 Summary of linear regression models of frequency traffic

Mode	β_0	β_1	R
1	0.1235	-1.0246×10^{-5}	0.020
2	0.1798	-1.5710×10^{-4}	0.161
3	0.2788	-7.6592×10^{-5}	0.069
4	0.3403	-1.3488×10^{-4}	0.142
5	0.3776	-1.0712×10^{-4}	0.062
6	0.4527	-7.1892×10^{-5}	0.056

Table 4.6 Variations of frequency induced by traffic loadings

Mode	Frequency at RMS 0 cm/s^2 (Hz)	Frequency at RMS 3 cm/s^2 (Hz)	Relative variation (%)
1	0.1235	0.1235	0.025
2	0.1798	0.1793	0.262
3	0.2788	0.2786	0.082
4	0.3403	0.3399	0.119
5	0.3776	0.3773	0.085
6	0.4527	0.4525	0.048

the correlation coefficient, R, between the measured RMS and normalized frequency sequences is obtained by

$$R = \frac{S_{fM}}{\sqrt{S_{ff}} \cdot \sqrt{S_{MM}}} \tag{4.6}$$

where S_{ff} is the variance of the normalized frequency sequence.

Table 4.5 summarizes the expressions of linear regression functions and the correlation coefficients of the modal frequency versus acceleration RMS. It is observed that all the regression functions have a negative slope, showing that the modal frequency slightly decreases with acceleration RMS in a statistical sense. Based on the regression functions, it is observed that the maximum and averaged relative variations of frequency induced by traffic are 0.262 and 0.104% for the variation of acceleration RMS covering a full cycle of in-service conditions, as illustrated in Table 4.6. The correlation coefficients are also less than 0.17 for all the six modes. Thus, it is concluded that the correlation between modal frequency and traffic loadings is very week and of no significance.

4.4.2 Correlation Analysis of Wind–Frequency

Correlation analysis is conducted on the normalized modal frequencies and the measured wind speeds under weak wind and strong wind conditions. Before the normalized model frequencies are used for wind–frequency correlation analysis, the traffic effect on the normalized modal frequencies should be removed. It is achieved by normalizing all the normalized modal frequencies to a fixed reference acceleration RMS with the use of the established regression functions of RMS frequency, as shown in Eq. (4.3). Similar to Eq. (4.2), the re-normalized modal frequencies after removing traffic effect are obtained by

$$f' = f - (f_V - f_{VR}) \tag{4.7}$$

where f' is the re-normalized modal frequency; f is the normalized modal frequency after temperature effect eliminating; f_V is the traffic-induced modal frequency and f_{VR} is the nominal modal frequency with the reference RMS 1 cm/s^2.

The re-normalized modal frequencies of all six modes at 10-min intervals and 10-min averaged wind speed measured at deck level are used in this study. Figure 4.12 shows the correlation diagrams of re-normalized frequency versus wind speed for the first, third, fifth, and sixth modes, respectively. It can be also observed that the wind speed–frequency plotting is rather scattered, and no apparent pattern exists. A linear regression analysis between the normalized frequency, f, and the wind speed, U, is performed by assuming [12, 13]

$$f' = \alpha_0 + \alpha_1 \times U \tag{4.8}$$

where the regression coefficients α_0 and α_1 are obtained by the least squares method.

Table 4.7 summarizes the expressions of linear regression functions and the correlation coefficients of the modal frequency versus wind speed. It is observed that all the regression functions have a positive slope, showing that the modal frequency slightly increases with wind speed in a statistical sense. Based on the regression functions, it is observed that the maximum and averaged relative variations of frequency induced by wind is 0.215 and 0.163% for the variation of wind speed covering a full cycle of in-service conditions, as illustrated in Table 4.8. The correlation coefficients are also less than 0.14 for all the six modes. Thus, it is also concluded that the correlation between modal frequency and wind speed is very week and of no significance.

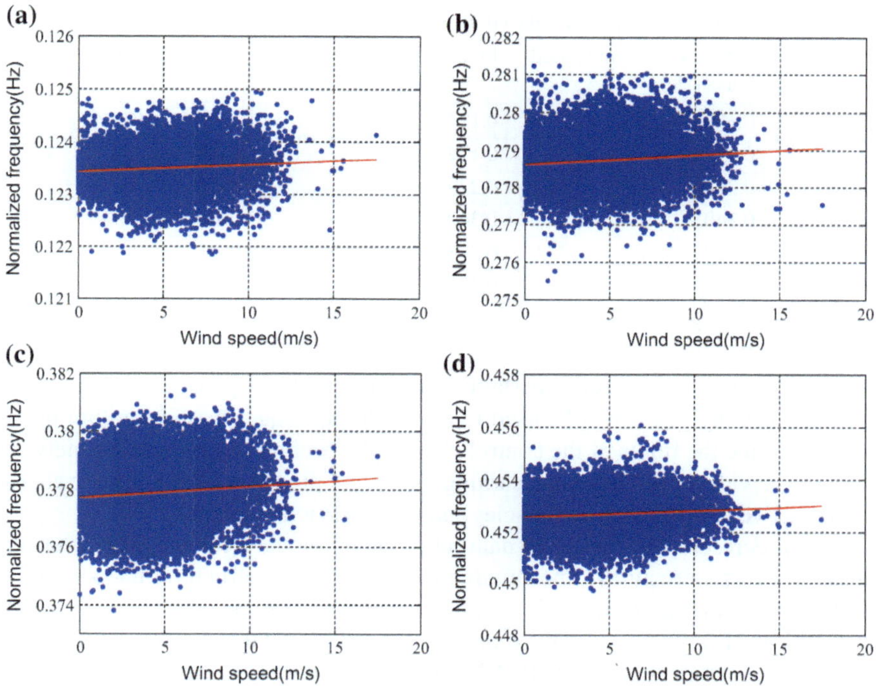

Fig. 4.12 Correlation between normalized modal frequency and wind speed: **a** Mode 1, **b** Mode 3, **c** Mode 4, and **d** Mode 6

Table 4.7 Summary of linear regression models of frequency wind

Mode	α_0	α_1	R (correlation coefficient)
1	0.1234	1.3281×10^{-5}	0.093
2	0.1796	0.7247×10^{-5}	0.030
3	0.2786	2.4866×10^{-5}	0.099
4	0.3401	3.3087×10^{-5}	0.135
5	0.3777	3.8282×10^{-5}	0.096
6	0.4526	2.3781×10^{-5}	0.083

4.5 Framework of Damage Detection

4.5.1 Machine Learning-Based Frequency–Temperature Model

The main purpose of the following framework is to detect the modal frequencies' abnormal changes caused by structural damages. The whole service period of the

Table 4.8 Variations of frequency induced by wind speed

Mode	Frequency at wind speed 0 m/s (Hz)	Frequency at wind speed 20 m/s (Hz)	Relative variation (%)
1	0.1234	0.1237	0.215
2	0.1796	0.1797	0.081
3	0.2786	0.2791	0.178
4	0.3401	0.3408	0.194
5	0.3777	0.3785	0.203
6	0.4526	0.4531	0.105

bridge structures is classified into healthy phase and unknown phase. The data coming from the healthy phase will be used to establish the frequency–temperature model and determine the limits of the control charts. The backpropagation neural network (BPNN), a widely used machine learning tool, is adopted to establish the correlation model between the modal frequencies and structural temperature.

Let f_i represents the measured data set of the ith modal frequency in the healthy phase of bridge structures. The simultaneously measured structural temperature is T. The length of both f_i and T is l. The temperature of the healthy phase T and the ith frequency f_i are used to be the training input vector and the training output vector of the BPNN, respectively. The BPNN-based T-f_i model is expressed as

$$f_i = \mathrm{BPF}_i(T) + e \tag{4.9}$$

where f_i represents the ith modal frequency. The independent variable T represents the structural temperature. e is the error. The temperature-eliminating frequency f_i^t is

$$f_i^t = f_i - \left[\mathrm{BPF}_i(T) - f_i^r\right] \tag{4.10}$$

$$f_i^r = \mathrm{BPF}_i\left(T^r\right) \tag{4.11}$$

where $\mathrm{BPF}_i(T)$ is the BPNN-based simulated frequency through substituting T into Eq. (4.9). T^r and f_i^r are the reference temperature and the ith reference frequency, respectively. Noting that the same reference temperature T^r will be used to eliminate the temperature effects of the modal frequencies from the unknown phase.

4.5.2 Probabilistically Modeling and Normalization

The X-bar control is introduced to identify the abnormal f_i^t. As a prerequisite, the variable is assumed to be normally distributed in the development of X-bar control charts [25]. This is not true in some cases. When the type of the underlying distribution is known, the sampling distribution can be accordingly derived. Some studies have focused on the effect of departures from normality on the control charts, which indicate that, in most cases, samples of size 4 or 5 are sufficient to ensure reasonable robustness to the normality assumption [25, 34]. However, the X-bar control chart is proposed to assess the individual modal frequencies. Thus, the sample size of the data set of each modal frequency is only one. In this situation, the non-normality of the modal frequencies should be addressed.

Herein, a method is proposed to address this problem. Before inputting f_i^t into the control chart, the hypothesis testing of normality is carried out. When f_i^t rejects the normal distribution, it should be transformed to a normal variable. According to the probability integral transform theorem, when random variable x has a continuous distribution function $F(x)$, $F(x)$ can be treated as a random variable U and U follows a (0, 1) uniform distribution. The concept of the Q statistic is introduced to address the non-normality of f_i^t [26]. The inverse transformation of the standard normal distribution is utilized to transform $F(x)$ into the Q statistic, which follows the standard normal distribution. First, it is necessary to find the distribution function $F(x)$. The kernel density estimate [35], which is a nonparametric density estimation technique, is adopted to obtain the distribution function of f_i^t. The kernel density estimate $\mathrm{PDF}(f_i^t)$ is

$$\mathrm{PDF}\left(f_{i,j}^t\right) = \frac{1}{lb}\sum_{k=1}^{l} K\left(\frac{f_{i,j}^t - f_{i,k}^t}{b}\right) \quad (j, k = 1, 2, \ldots, l) \tag{4.12}$$

where l is the data length mentioned in Sect. 4.5.1. $f_{i,j}^t$ is the jth data point of the temperature-elimination frequency f_i^t. The kernel function $K(\cdot)$ is the standard normal function in this study. After the kernel function is determined, the bandwidth b can be selected by using the optimization method [36]. In this study, the method of the cross-validation [36] is employed to obtain the optimal bandwidth b. Then, the cumulative probability estimate $\mathrm{CDF}(f_i^t)$ is

$$\mathrm{CDF}\left(f_{i,j}^t\right) = \frac{1}{lb}\sum_{k=1}^{l} \int_{-\infty}^{f_{i,j}^t} K\left(\frac{x - f_{i,k}^t}{b}\right)\mathrm{d}x \quad (j, k = 1, 2, \ldots, l) \tag{4.13}$$

$\mathrm{CDF}(f_i^t)$ can be transformed into the Q statistic as

$$Q_i = \Phi^{-1}\left[CDF\left(f_i^t\right)\right] \tag{4.14}$$

where Q_i is the Q statistic of the ith modal frequency. $\Phi^{-1}(\cdot)$ is the inverse cumulative distribution function of the standard normal distribution.

4.5.3 Control Charts of the Healthy Phase

The X-bar control chart provides a technique for monitoring the changes of the selected feature means and identifies observation points that are inconsistent with the previous data sets. The main theory using the X-bar control chart to monitor the mean variation of the features is introduced as follows [37]. Suppose that variable X is normally distributed with mean μ and standard deviation σ, where μ and σ are known. When the size of sample x_1, x_2, ..., x_n is n, the mean value of the sample is $\bar{x}$. $\bar{x}$ is also normally distributed with mean μ and standard deviation $\sigma/\sqrt{n}$. The probability of $\bar{x}$ can be expressed as

$$P\left(\left|\frac{\bar{x}-\mu}{\sigma/\sqrt{n}}\right| < Z_{\alpha/2}\right) = 1-\alpha \tag{4.15}$$

where α is the significance level and $Z_{\alpha/2}$ is the percentage point of the standard normal distribution such that $P\left(Z \geq Z_{\alpha/2}\right) = \alpha/2$. The sample mean will be within the range of $\left(\mu - Z_{\alpha/2}\frac{\sigma}{\sqrt{n}}, \mu + Z_{\alpha/2}\frac{\sigma}{\sqrt{n}}\right)$.

In this study, the bridge modal frequencies are assessed individually. For the normally distributed variable Q_i of the ith modal frequency, the formulas for constructing the control limits on the X-bar control chart are

$$\mathrm{UCL} = \mathrm{mean}(Q_i) + Z_{\alpha/2}\frac{\mathrm{std}(Q_i)}{\sqrt{n}} \tag{4.16a}$$

$$\mathrm{CL} = \mathrm{mean}(Q_i) \tag{4.16b}$$

$$\mathrm{LCL} = \mathrm{mean}(Q_i) - Z_{\alpha/2}\frac{\mathrm{std}(Q_i)}{\sqrt{n}} \tag{4.16c}$$

where mean(Q_i) and std(Q_i) represents the mean value and standard deviation of Q_i, respectively.

4.5.4 Control Charts of the Unknown Phase

Let Uf_i represent the ith measured modal frequency within unknown phase. The corresponding structural temperature is UT. The lengths of both Uf_i and UT are m. First, UT is put into BPNN-based correlation model (Eq. 4.9). Then, Eq. (4.10) is used to eliminate the temperature effect on Uf_i

$$Uf_i^t = Uf_i - \left[\mathrm{BPF}_i(UT) - f_i^r\right] \tag{4.17}$$

where Uf_i^t is the temperature-eliminating ith frequency in unknown phase. The kernel density estimate $\mathrm{PDF}\left(Uf_i^t\right)$ is

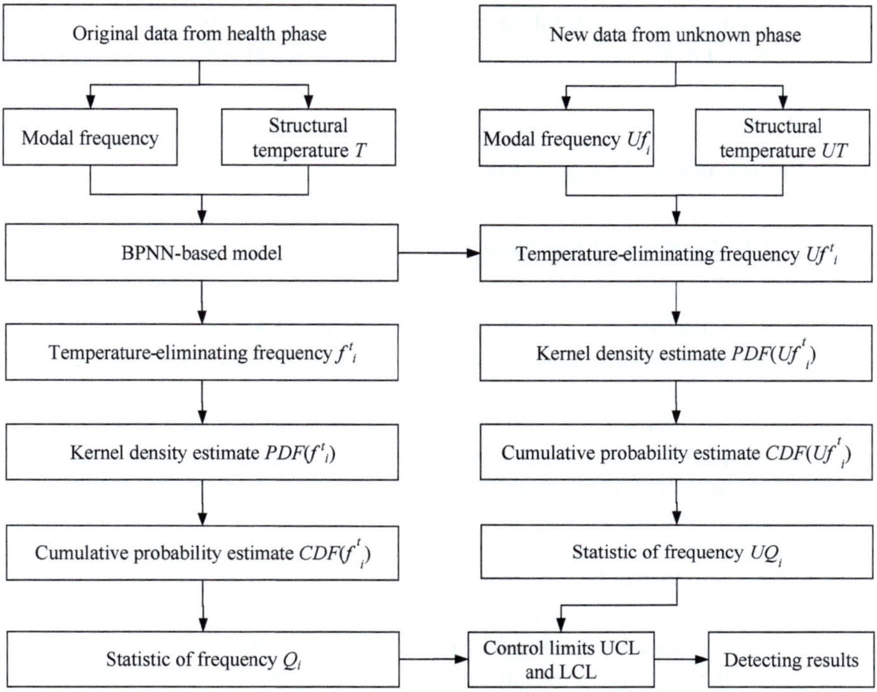

Fig. 4.13 Flowchart of the probabilistic monitoring-based framework

$$\mathrm{PDF}\left(Uf_{i,s}^{t}\right) = \frac{1}{lb}\sum_{k=1}^{l} K\left(\frac{Uf_{i,s}^{t} - f_{i,k}^{t}}{b}\right) \quad (k = 1, 2, \ldots, l; s = 1, 2, \ldots, m) \quad (4.18)$$

where l and m are the lengths of the data sets in healthy and unknown phase, respectively. $Uf_{i,s}^{t}$ is the sth data point of Uf_i^t. Furthermore, the cumulative probability estimate $\mathrm{CDF}\left(Uf_i^t\right)$ can be obtained as

$$\mathrm{CDF}\left(Uf_{i,s}^{t}\right) = \frac{1}{lb}\sum_{k=1}^{l} \int_{-\infty}^{Uf_{i,s}^{t}} K\left(\frac{x - f_{i,k}^{t}}{b}\right)\mathrm{d}x \quad (k = 1, 2, \ldots, l; s = 1, 2, \ldots, m) \quad (4.19)$$

$\mathrm{CDF}\left(Uf_i^t\right)$ is also transformed into the Q statistic

$$UQ_i = \Phi^{-1}\left[\mathrm{CDF}\left(Uf_i^t\right)\right] \quad (4.20)$$

where UQ_i are the Q statistics of the ith modal frequency of the unknown phase. Finally, UQ_i will be input to the control chart established in Sect. 4.5.3. The data point which exceeds the control limits indicates the abnormal change of the modal frequency induced by structural damage. Figure 4.13 shows the flow chart of the proposed probabilistic assessment framework for long-span bridges.

4.6 Framework Application

4.6.1 Elimination of the Temperature Effects

The modal frequencies are identified per 10 minutes. Then for every 6 hours, 36 identified frequencies are averaged to a value to establish the frequency data set. Hence, there are four data points for each natural frequency in each day. The temperature data from all sensors of the mid-span section are averaged to represent the structural temperature field of the RSB. The length of data sets of frequencies and temperature are both 860. All data should be considered to be from the healthy phase since it was a short time since the RSB opened to traffic. To illustrate the assessment of the modal frequencies, the original measurements are divided into two subsets. Data points from 301 to 350 are selected from the original measurements and will be used as the monitoring data for the unknown phase of the RSB. Thus, the other 810 data points formed the monitoring data set for the healthy phase. Tables 4.9 and 4.10 tabulate the statistical characteristics of modal frequencies from both healthy and unknown phases. Note that, the mode shapes illustrated in Table 4.9 are obtained by the finite element analysis.

Because higher order modal frequencies are more sensitive to damages, the fourth, fifth and sixth vertical bending modal frequencies are selected to illustrate the framework presented in this study. Figure 4.14 presents the evolution of the three modal frequencies and structural temperature during the period under analysis. It could be observed in this figure that all measured frequencies have almost the same variation pattern but diametrically opposed to that of the structural temperature.

Section 4.6.1 aims to present the procedures to establish the correlation model to eliminate (or minimize) the effects caused by the temperature. The data from the healthy phase will be used to establish the temperature–frequency regression model and then the data collected during the unknown phase will be used to validate its quality from the forecasts provided by the models. The f_i-T points are plotted as solid dots in scatter graphs of Fig. 4.15. For each frequency, an overall decrease in the modal frequency is observed with the increase of structural temperature. However, the f_i-T relationships expressed in these figures are nonlinear. Therefore, the linear regression models are not suitable to establish the correlation model. Herein, the backpropagation neural network is adopted to conduct the correlation analysis. The BPNN training is carried out by using the LM algorithm with the Bayesian regularization technique, which can effectively improve the BPNNs performance by addressing the over-fitting problem [15, 38]. Figure 4.15 shows the structure of the single hidden layer backpropagation neural network. The training input and output of the network are the structural temperature T and the modal frequency f_i, respectively.

Table 4.9 Summary of statistics of the measured modal frequencies f_i of the RSB

Mode order	Maximum (Hz)	Minimum (Hz)	Mean value (Hz)	Relative variation (%)	Standard deviation (Hz)	Mode shape and description
f_1	0.1242	0.1232	0.1235	0.81	1.42e−04	 First symmetric vertical mode
f_2	0.1811	0.1789	0.1797	1.22	3.21e−04	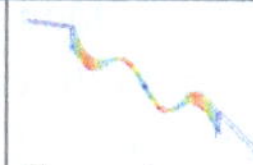 Second anti-symmetric vertical mode
f_3	0.2816	0.2778	0.2790	1.36	6.56e−04	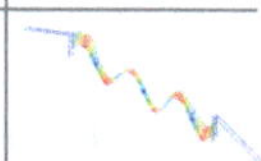 Third anti-symmetric vertical mode
f_4	0.3452	0.3390	0.3408	1.82	1.30e−03	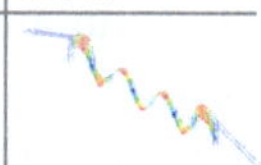 Fourth symmetric vertical mode
f_5	0.3836	0.3748	0.3783	2.33	2.60e−03	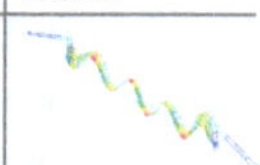 Fourth anti-symmetric vertical mode
f_6	0.4600	0.4508	0.4535	2.03	2.2e−03	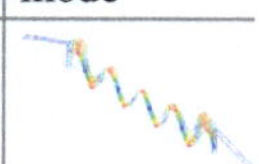 Fifth symmetric vertical mode

Table 4.10 Summary of statistics of the measured modal frequencies Uf_i of the RSB

Mode order	Maximum (Hz)	Minimum (Hz)	Mean value (Hz)	Relative variation (%)	Standard deviation (Hz)
Uf_4	0.3417	0.3399	0.3406	0.53	4.95e−04
Uf_5	0.3809	0.3763	0.3787	1.21	1.80e−03
Uf_6	0.4551	0.4519	0.4533	0.71	9.46e−04

Note Relative variation = (Maximum-Minimum)/mean value × 100%

The hidden node number is critical to the generalization capability of the network. Herein, to ensure the best prediction capability for the trained networks, the coefficient of the mean relative error (MRE) is used to determine the optimal number of the hidden node

$$\mathrm{MRE} = \frac{1}{m}\sum_{j=1}^{m}\left|\frac{\mathrm{BPF}_i(UT_j) - Uf_{i,j}}{Uf_{i,j}}\right| \tag{4.21}$$

where m, mentioned previously in Sect. 4.5.4, is the length of the testing data from the unknown phase. $Uf_{i,j}$ and UT_j are the jth data points of the ith modal frequency and structural temperature, respectively.

The number of hidden nodes of the networks is assumed to vary between 1 and 20. With a certain hidden node number, T and f_i are used to train the network for each modal frequency. Then, UT is fed to the trained network, and MRE of each hidden node number can be obtained. Figure 4.16 shows the prediction capability of the networks in terms of the number of hidden nodes. When the hidden node number is 2, MREs of the three modal frequencies achieve the minimum values, i.e., 0.041, 0.087, and 0.053%, respectively. Therefore, the optimal number of hidden nodes is set to 2. Due to the random initialization of the network parameters during each training process, same training data and structure can result in different networks. Thus, the networks with the optimal number of hidden nodes will be saved and used in the following analysis.

Figure 4.17 further shows the simulated curves of temperature and modal frequencies by feeding temperature into the trained BPNNs. It is clear that the simulated curves can characterize the relationships between structural temperature and modal frequencies. UT is also fed into the trained BPNNs of the selected three modal frequencies. Then the original Uf_i and predicted values are plotted in Fig. 4.17. It could be observed that the predicted values are in satisfactory agreement with the original Uf_i. The results in this figure indicate that BPNN, as a nonlinear mapping tool with strong fault tolerance, shows strong generalization capability in modeling the correlations between structural temperature and modal frequencies.

Equations (4.10) and (4.17) are used to obtain the temperature-eliminating modal frequencies f_i^t and Uf_i^t. Noteworthy that the reference temperature T^r is 20 °C. For example, Fig. 4.18 compares the original and temperature-eliminating sixth-

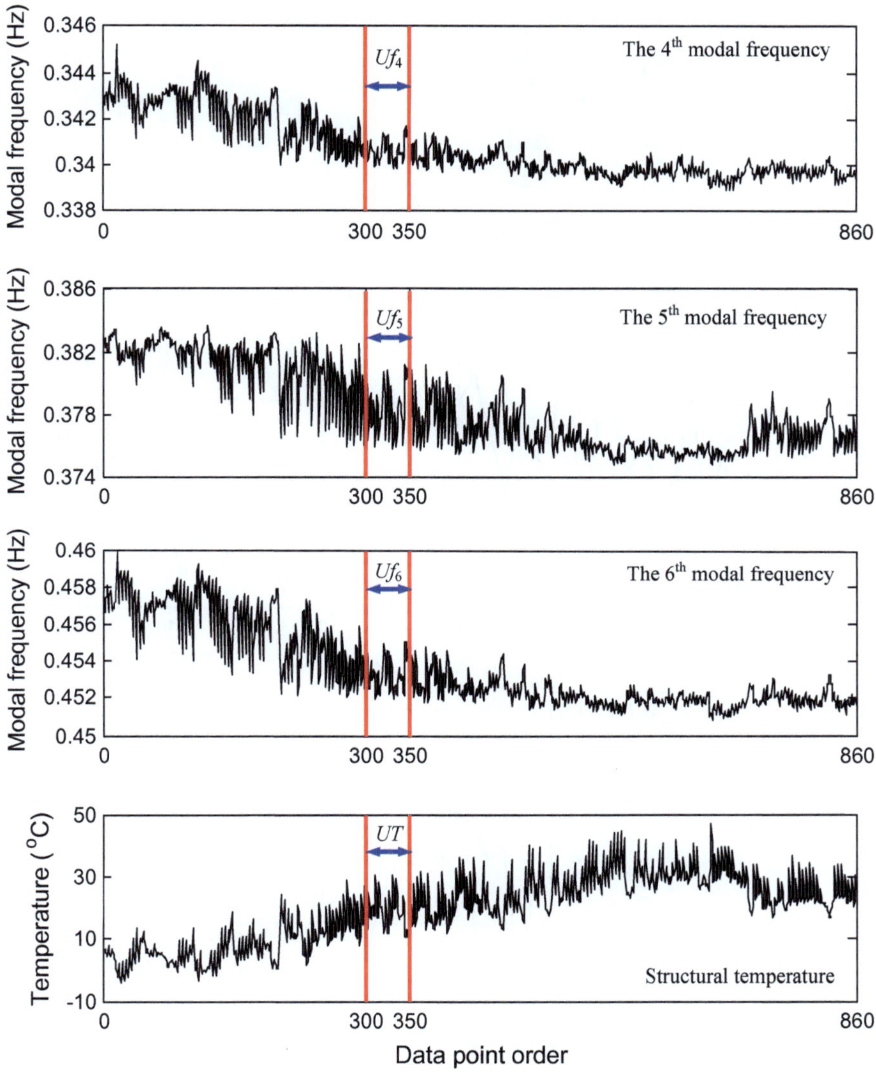

Fig. 4.14 Time evolution of the bridge frequencies versus structural temperature

order modal frequencies. It is observed that the trend component induced by the temperature changing is eliminated from the measured modal frequency.

The reduction of the variation of the modal frequencies after the temperature elimination by the BPNNs can be validated by histograms. Figure 4.19 presents the histograms of the modal frequencies. It is obvious that the frequencies are more concentrated in a narrower range with one single peak after elimination of the temperature effects.

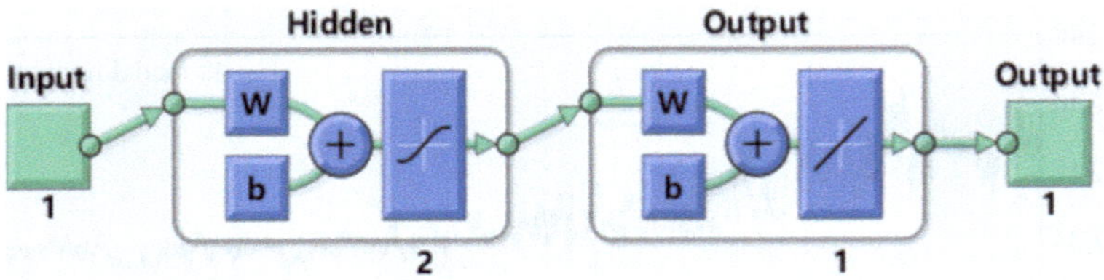

Fig. 4.15 Structure of the single hidden layer neural network

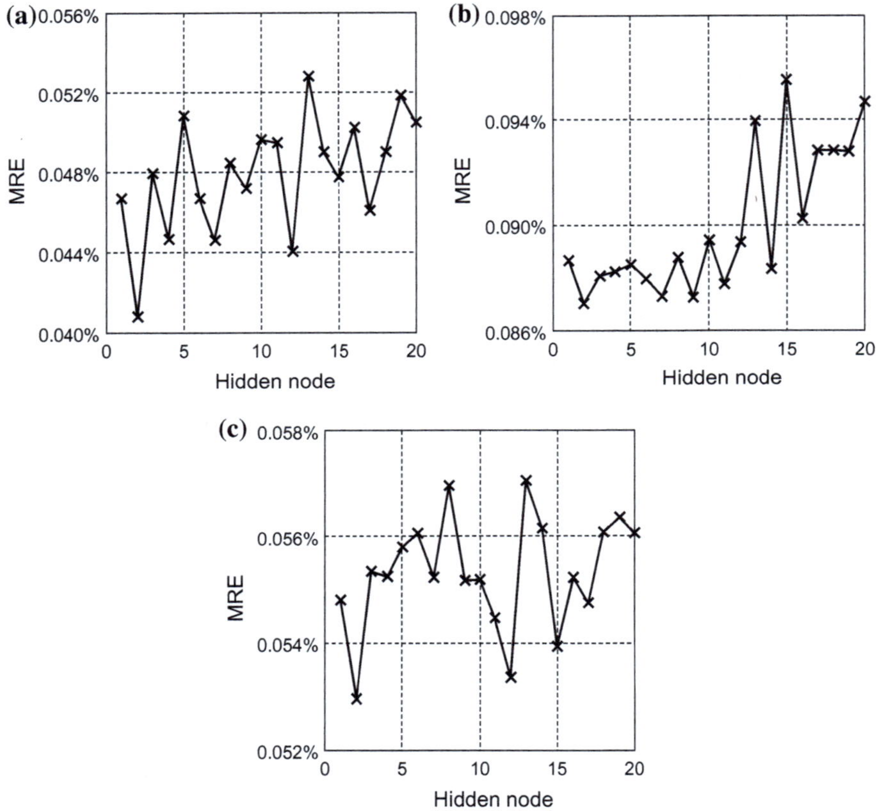

Fig. 4.16 Mean relative error of the networks for: **a** f_4, **b** f_5, and **c** f_6

The histograms of f_i^t also show that besides the temperature, there are other factors, including the vehicle loading and wind action, that may also result in variations of the modal frequencies. The randomness in the histograms of f_i^t does not allow the direct identification of the abnormal changes of the bridge modal frequencies. In other words, f_i^t of each modal frequency should be a constant if the model frequencies of bridges are only influenced by temperature under the healthy phase. Therefore, if efforts can be made to make f_i^t be a constant, it would be much easier to conduct

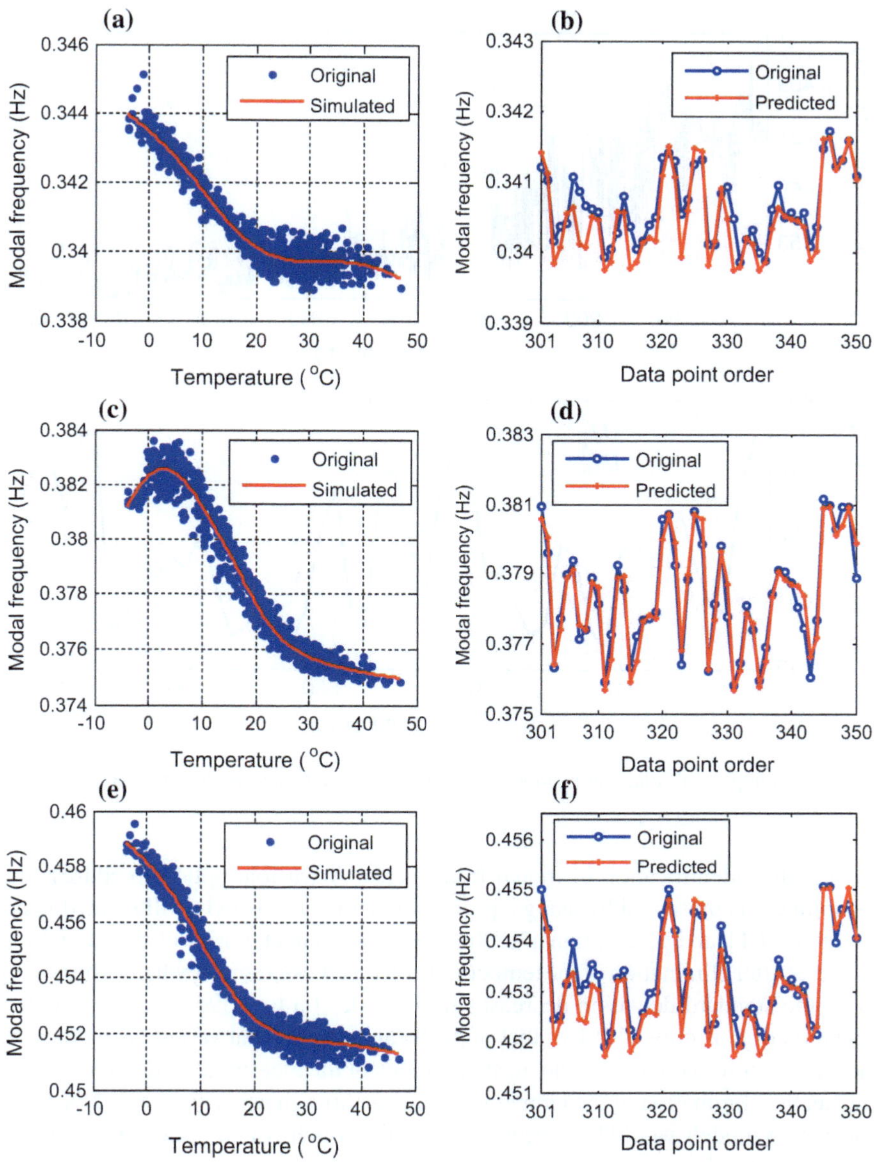

Fig. 4.17 Correlation models between modal frequencies and structural temperature. **a** f_4 versus T, **b** original and predicted values of Uf_4, **c** f_5 versus T, **d** original and predicted values of Uf_5, **e** f_6 versus T, and **f** original and predicted values of Uf_6

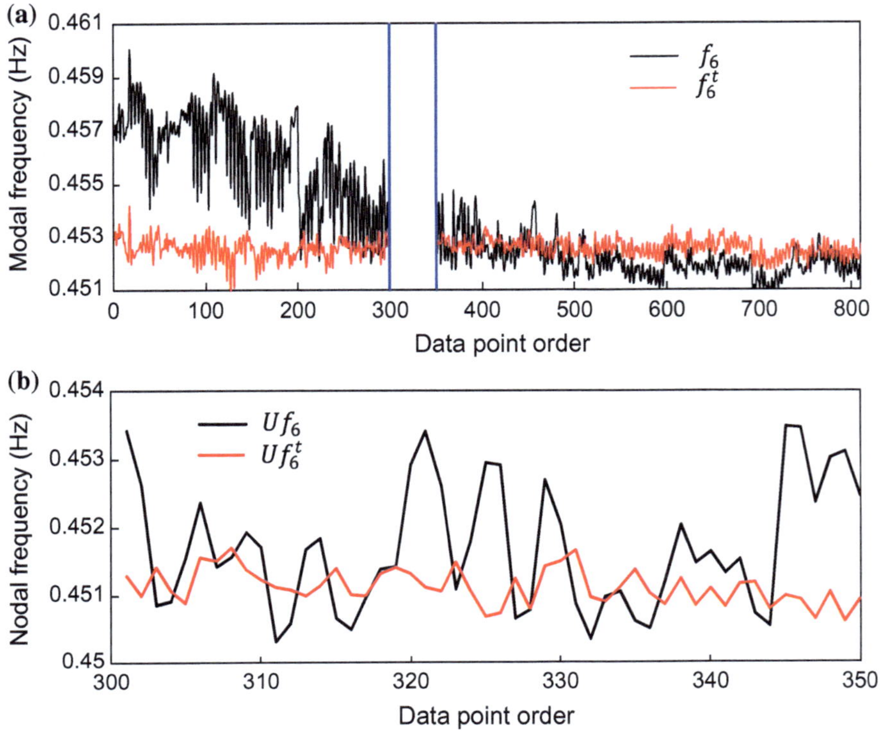

Fig. 4.18 Temperature-eliminating modal frequency: **a** training data and **b** testing data

abnormality identification of modal frequencies. A feasible option to address this randomness is presented by using a probabilistic method, which is illustrated later.

Tables 4.11 and 4.12 summarize the statistical characteristics of the modal frequencies after elimination of the temperature effects. For example, when f_5 becoming to f_5^t, the standard deviation decreases from 2.60e−03 to 4.39e−04. The standard deviation of f_5^t is only 16.9% of that of f_5. It can be noted that variations induced by the temperature accounts for the most variations of the measured modal frequencies. Hence, the elimination of the temperature effects through applying the BPNN-based correlation models would be expected to effectively help identify the abnormality of modal frequencies.

Besides, the histograms of f_i^t in Fig. 4.19 seem to follow normal distribution. If f_i^t can be proved to be normally distributed, the X-bar control chart could be directly used to identify the abnormality of modal frequencies of the unknown phase. Thus, four kinds of hypothesis testing are employed. As shown in Table 4.13, the results show that the modal frequencies after elimination of the temperature effects reject normal distribution. Therefore, after the temperature elimination, the normalization of modal frequencies should be further carried out.

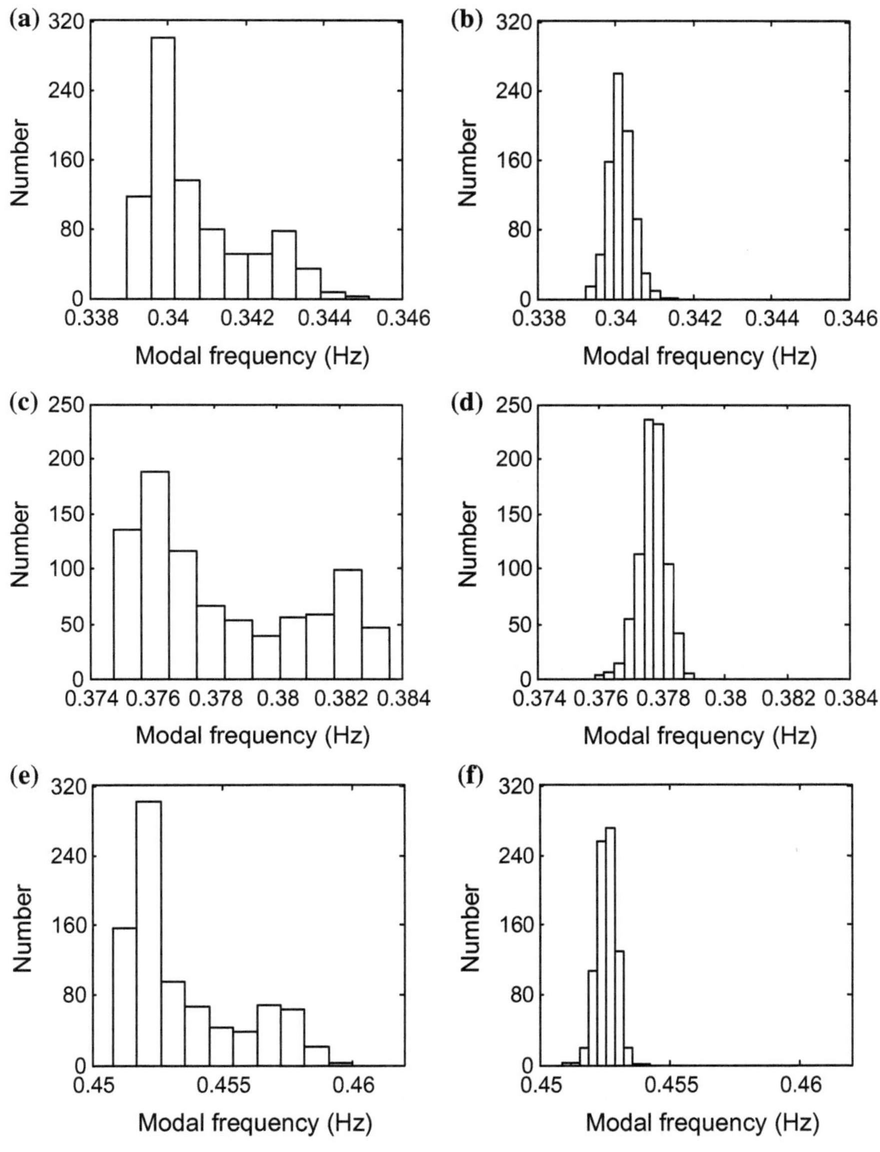

Fig. 4.19 Histograms of the modal frequencies (from the fourth to sixth) before and after the elimination of the temperature effects: **a** f_4, **b** f_4^t, **c** f_5, **d** f_5^t, **e** f_6, and **f** f_6^t

4.6.2 *Normalization of the Modal Frequencies*

Section 4.6.2 describes the normalization processes of the modal frequencies after the temperature elimination. Taking the sixth modal frequency as an example, Fig. 4.20

Table 4.11 Summary of statistics of modal frequencies f_i^t after elimination of the temperature effects

Mode order	Maximum (Hz)	Minimum (Hz)	Mean value (Hz)	Relative variation (%)	Standard deviation (Hz)
f_4^t	0.3416	0.3393	0.3401	0.68	3.07e−04
f_5^t	0.3790	0.3759	0.3777	0.85	4.39e−04
f_6^t	0.4542	0.4509	0.4526	0.73	3.62e−04

Table 4.12 Summary of statistics of modal frequencies Uf_i^t after elimination of the temperature effects

Mode order	Maximum (Hz)	Minimum (Hz)	Mean value (Hz)	Relative variation (%)	Standard deviation (Hz)
Uf_4^t	0.3409	0.3399	0.3403	0.29	2.43e−04
Uf_5^t	0.3784	0.3765	0.3776	0.50	4.32e−04
Uf_6^t	0.4533	0.4523	0.4528	0.22	2.62e−04

Note Relative variation = (Maximum-Minimum)/mean value × 100%

Table 4.13 Results of hypothesis test of normal distribution for f_i^t

Modal frequencies after elimination of the temperature effects		f_4^t	f_5^t	f_6^t
χ^2 goodness-of-fit testing	h	1	1	1
	p	0.0029	5.16e−004	0.003
Kolmogorov–Smirnov testing	h	1	1	1
	p	5.32e−134	1.73e−296	3.85e−322
Lilliefors testing	h	1	1	1
	p	0.0009	0.001	0.0084
Jarque–Bera testing	h	1	1	1
	p	0.001	0.001	0.001

presents the normalization process. First, the kernel density estimation is used to obtain the probability density function of f_6^t. Then $\mathrm{PDF}(f_6^t)$ and $\mathrm{CDF}(f_6^t)$ are derived by using Eqs. (4.12) and (4.13), respectively. At last, the Q statistic Q_6 is calculated from $\mathrm{CDF}(f_6^t)$ by using Eq. (4.14). Results of four kinds of hypothesis testing are presented in Table 4.14. As expected, Q_4, Q_5, and Q_6 are now considered to be normally distributed, as shown in Fig. 4.21.

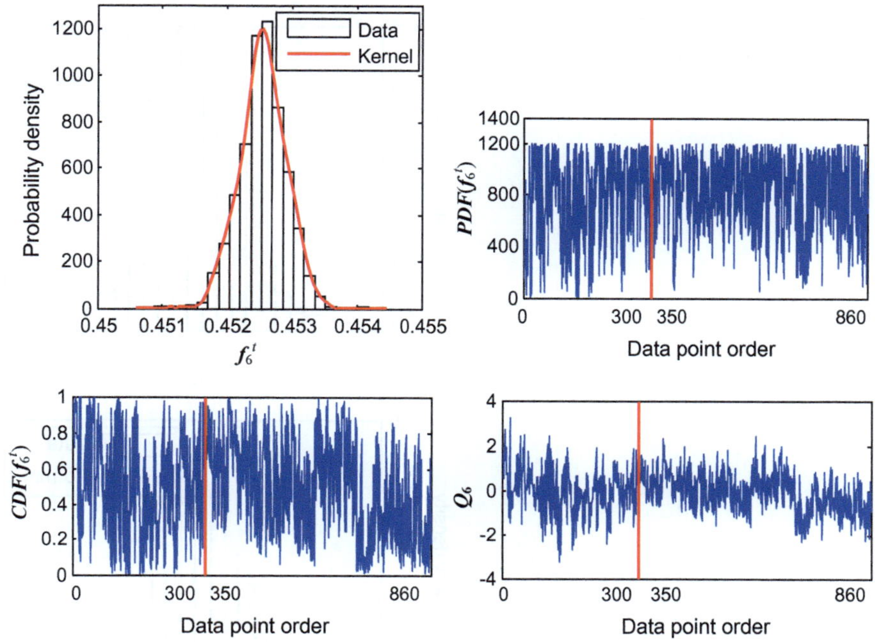

Fig. 4.20 Normalization process of the sixth modal frequency

Table 4.14 Results of hypothesis test of normal distribution for Q_i

Q statistics of modal frequencies		Q_4	Q_5	Q_6
χ^2 goodness-of-fit testing	h	0	0	0
	p	0.84	0.72	0.99
Kolmogorov–Smirnov testing	h	0	0	0
	p	0.78	0.73	0.72
Lilliefors testing	h	0	0	0
	p	0.50	0.50	0.50
Jarque–Bera testing	h	0	0	0
	p	0.50	0.50	0.50

Note The significance level of hypothesis testing in Tables 4.13 and 4.14 is 0.05; the coefficients h and p are the results of hypothesis testing; if $h = 0$ and $p > 0.05$, the variable is normally distributed; if $h = 1$ and $p < 0.05$, it means that the variable rejects normal distribution

4.6.3 Damage Detection Based on Control Chart

The control limits of each modal frequency are established based on the data from the healthy phase. The new data from the unknown phase is fed into the existing control charts to check whether there are abnormal changes in the modal frequencies,

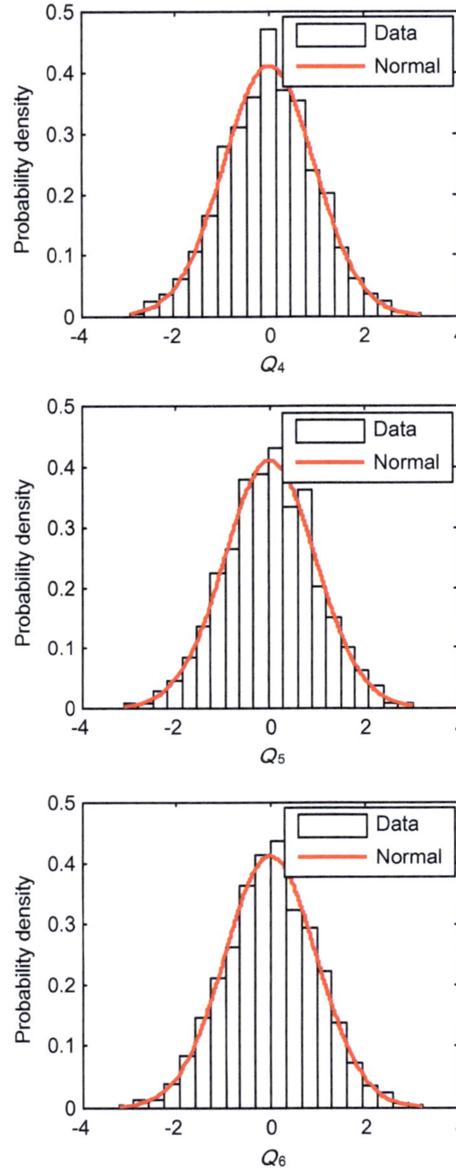

Fig. 4.21 Fitting results of probability density function of Q_i

which may be induced by structural damage or extreme loading conditions and can't be identified through simple comparison with the data from the healthy phase. In addition, Eqs. (4.17)–(4.20) are used to transfer Uf_4, Uf_5, and Uf_6 into Q statistics UQ_4, UQ_5, and UQ_6.

Table 4.15 Summary of statistical characteristics of Q_i

Q_i	Q_4	Q_5	Q_6
Mean value	0.0014	−0.002	-8.615×10^{-4}
Standard deviation	0.97	0.97	0.97

When the mean value and standard deviation of Q_i are determined (as shown in Table 4.15), the significance level α needs to be studied to calculate the UCL and LCL according to Eq. (4.16). Some studies proposed that the significance level could be 0.27% ($Z_{\alpha/2} = 3.0$) [16, 17, 19]. It suggests that the value of significance level should be determined to reduce the risk of errors of the control charts. Two kinds of errors may be committed when feeding data into the control charts [25]. If some data points exceed the control limits when all data points are in normal state, then a type I error or false alarm has occurred. If there are no data points exceed the control limits when some data points are in abnormal state, then a type II error has been made. In this study, the data of both the health and unknown phases are measured within the health state of the RSB, so the determination of significance level α should focus on reduction risk of type I error.

Figure 4.22 shows the determination process of significance level. When the significance level α changes from 0.27 to 0.10%, there are points exceeding the control limits of the training data set. The points that exceed the control limits mean type I error. When α equals to 0.06% ($Z_{\alpha/2} = 3.4$), the range between UCL and LCL gets wider and both Q_i of the healthy phase and UQ_i of the unknown phase are within the limits of the X-bar charts, shown in Fig. 4.23.

On the other hand, the occurrence probability of type II error implies the sensitivity of abnormal data of the control charts. Higher sensitivity means lower occurrence probability of the type II error and poorer sensitivity will result in larger probability of the type II error. The sensitivity analysis of the control charts presented in Fig. 4.24 are carried out to evaluate the occurrence probability of the type II error. Since the RSB is a relatively new bridge, damage scenarios representing the unknown phase are numerically simulated in this study. Damage of suspension bridges may be resulted from many sources, e.g., cracks in bridge tower due to an earthquake, cross-sectional area loss of main cable cause by corrosion, fracture of hangers induced by the coupling of fatigue and corrosion, etc. The exact modeling of damage scenarios requires complex nonlinear models and time-consuming calculation, which is beyond the scope of this study. In most cases, damage would make the actual modal frequencies become lower. Hence, the ith modal frequency $Uf_{i,\delta}$ of the damage structure can be simulated as [39]

$$Uf_{i,\delta} = Uf_i - \delta \cdot \text{mean}(f_i) \tag{4.22}$$

where δ is the parameter representing the damage degree; mean(f_i) is the mean value of f_i. According to Table 4.9, mean(f_4), mean(f_5), and mean(f_6) are 0.341, 0.378,

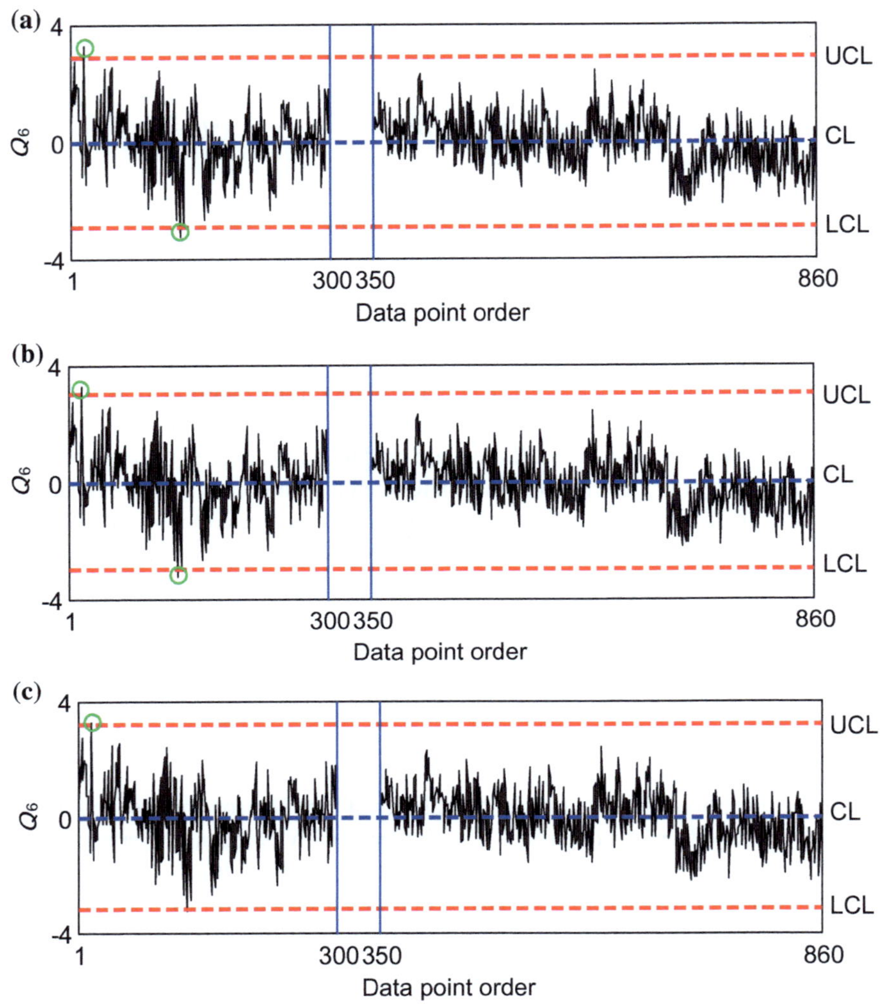

Fig. 4.22 Control charts of the training data: **a** $\alpha = 0.27\%$, **b** $\alpha = 0.20\%$, and **c** $\alpha = 0.10\%$

and 0.454 Hz, respectively. Equations (4.17)–(4.20) are used to transfer $Uf_{i,\delta}$ into $UQ_{i,\delta}$, and then $UQ_{i,\delta}$ is input into the same control charts shown in Fig. 4.23.

Figure 4.24 presents the control chart of the unknown phase, or the so called simulated damage state. It is observed that: (i) For the fourth modal frequency, there are 2 data points that exceed the LCL when δ equals to 0.2%. Then when δ increased to be 0.3%, there are 18 points outside LCL. The abnormal change of the 4th modal frequency caused by simulated structural damages (i.e., corresponding to $\delta = 0.2\%$ and 0.3%) can be detected. (ii) For the fifth modal frequency, when δ is 0.15%, all points of $UQ_{5,0.15\%}$ are still within the range between UCL and LCL. It implies that the control chart does not detect the abnormality of the fifth modal

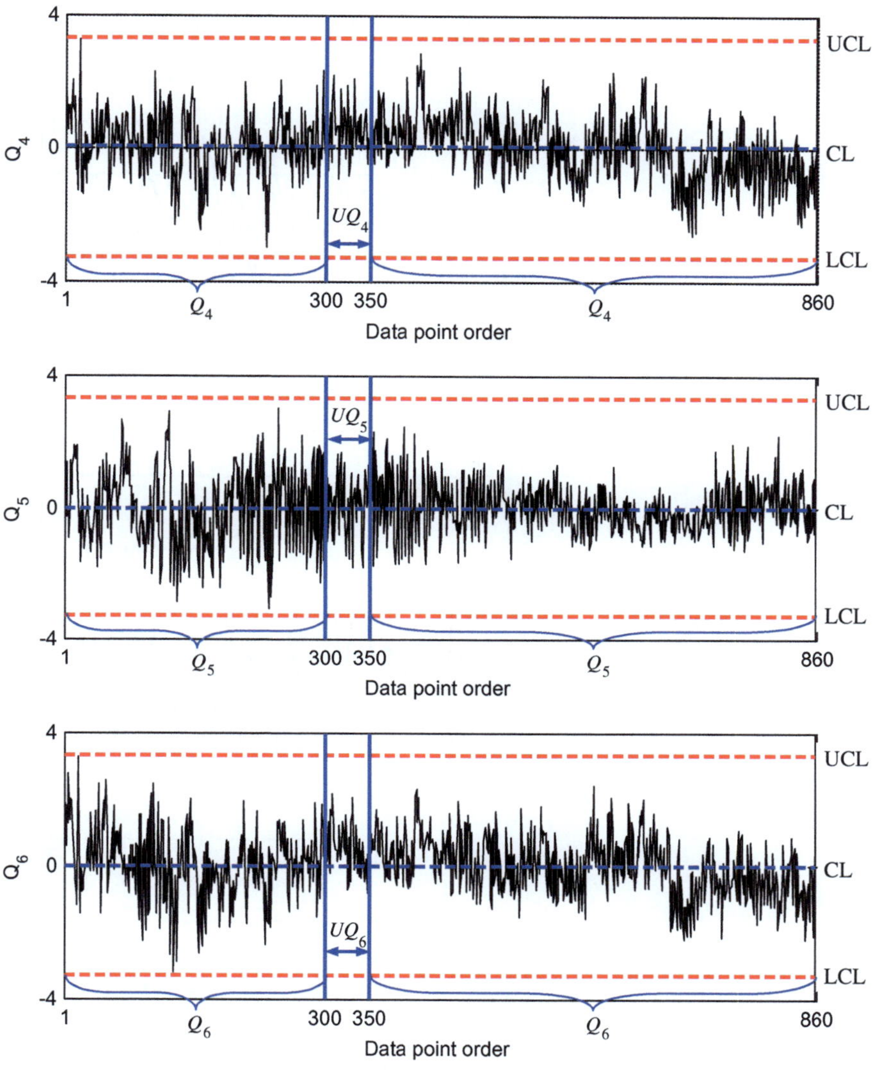

Fig. 4.23 Control charts in healthy phase

frequency and the type II error occurs. When δ increases to be 0.3%, the damage can be detected because of 5 points exceeding LCL. (iii) When δ becomes to be 0.35%, the abnormality of the sixth modal frequency has been detected with 6 points exceeding LCL. The above results indicate that only when the abnormal changes of modal frequencies caused by structural damage or some other reasons accumulates to a certain extent, the probabilistic monitoring-based framework can be implemented effectively. However, when the abnormal change is smaller, then the control chart cannot function properly, and the type II error is inevitable. After all, the results in

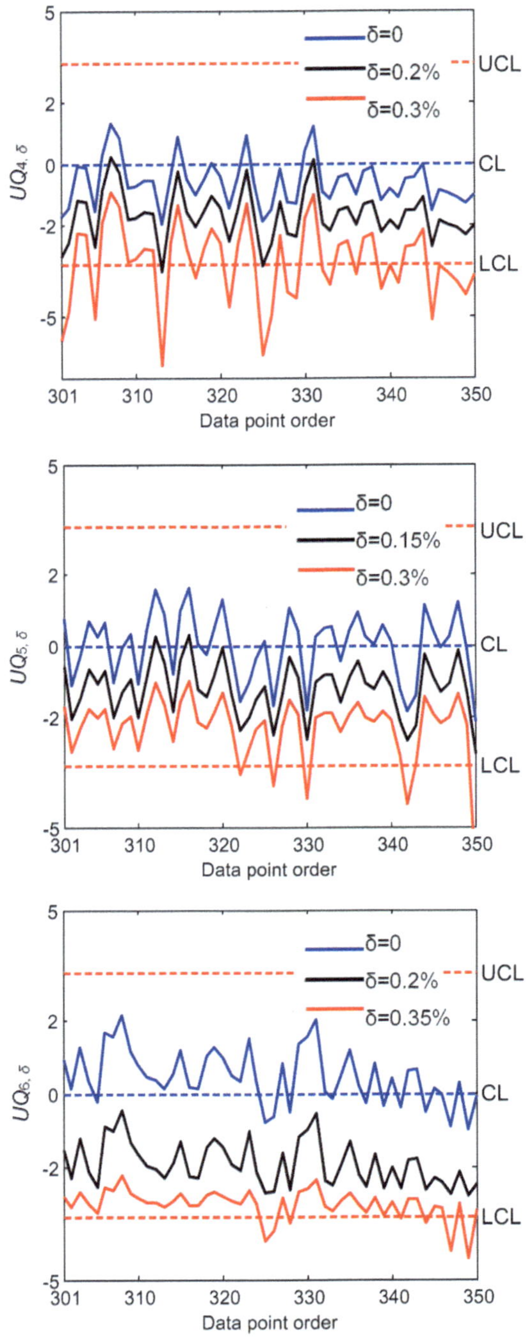

Fig. 4.24 Control charts in unknown phase

Fig. 4.24 suggest that the framework based on long-term monitoring and probability theory has shown its sensitivity and can be used to detect minor abnormalities of the modal frequencies of the RSB.

4.7 Summary

In this chapter, the variations of modal frequencies with normal environmental conditions have been studied for the RSB based on long-term continuous measurement data. The output-only modal frequency identification of the bridge has been effectively carried out from 10-month measurement data using the iterative windowed curve-fitting method (IWCM) in the frequency domain. Furthermore, a new probabilistic monitoring-based framework for damage detection of long-span bridges utilizing the modal frequencies is proposed. The following are the findings and conclusions.

1. From the identified frequency sequences in a typical day, it can be observed that the measured frequencies present instantaneous changes because of the nonstationary properties of the ambient loadings. In order to eliminate the random variations rising from the identification algorithm, the daily averaged frequencies using multi-sample averaging technique were obtained, i.e., the seasonal correlation analysis of frequency–temperature. The maximum and averaged relative variations of modal frequencies reduced from 3.013 and 2.633% measured at 10-min intervals to 2.168% and 1.268% using multi-sample averaging technique.
2. Temperature is the critical source causing modal variability, and there is an overall decrease in modal frequency with temperature for all the identified modes. A 6-order polynomial regression model is further applied for the modeling of daily averaged frequency and temperature. It is found that the maximum and averaged relative variations of frequency induced by temperature are 1.975 and 1.175% for the variation of temperature covering a full cycle of in-service conditions.
3. The measured variations in modal frequencies mainly arise from temperature, wind speed, traffic loadings, and identification algorithm. The modal variability caused by wind speed and traffic loadings is notably less than by temperature and identification algorithm. And the influence of identification algorithm on modal frequencies is as significant as the temperature effect. Thus, for reliable performance of vibration-based damage detection methods, the improvement of the modal frequency identification algorithm properly considering the nonstationary properties of the ambient loadings is very important. Based on the accuracy measurement of modal frequencies, the normal variations induced by various environmental factors can then be quantified and further serve for structural health monitoring and damage alarming.
4. The proposed framework is demonstrated on a prototype long-span suspension bridge and a 215-day monitored data are analyzed. The framework can address the issue of non-normal distribution of the input data when adopting the control

charts to detect damage using the modal frequencies. The results show that this framework had good sensitivity to the modal frequencies' abnormality and could be applied to the SHM of long-span bridges.

References

1. Ko JM, Ni YQ. Technology developments in structural health monitoring of large-scale bridges. Eng Struct. 2005;27(12):1715–25.
2. Hsieh KH, Halling MW, Barr PJ. Overview of vibrational structural health monitoring with representative case studies. J Bridge Eng. 2006;11(6):707–15.
3. Ni YQ, Hua XG, Fan KQ, Ko JM. Correlating modal properties with temperature using long-term monitoring data and support vector machine technique. Eng Struct. 2005;27(12):1762–73.
4. Ding YL, Li AQ, Liu T. Environmental variability study on the measured responses of Runyang Cable-stayed Bridge using wavelet packet analysis. Sci China Ser E: Technol Sci. 2008;51(5):517–28.
5. Cornwell P, Farrar CR, Doebling SW, Sohn H. Environmental variability of modal properties. Exp Tech. 1999;23(6):45–8.
6. Peeters B, De Roeck G. One-year monitoring of the Z24-Bridge: environmental effects versus damage events. Earthquake Eng Struct Dynam. 2001;30(2):149–71.
7. Sohn H, Dzwonczyk M, Straser EG, Kiremidjian AS, Law KH, Meng T. An experimental study of temperature effect on modal parameters of the Alamos Canyon Bridge. Earthquake Eng Struct Dynam. 1999;28(8):879–97.
8. Wahab AM, De Roeck G. Effect of temperature on dynamic system parameters of a highway bridge. Struct Eng Int. 1997;7(4):266–70.
9. Hua XG, Ni YQ, Ko JM, Wong KY. Modeling of temperature-frequency correlation using combined principal component analysis and support vector regression technique. J Comput Civil Eng. 2007;21(2):122–35.
10. Kim CY, Jung DS, Kim NS, Yoon JG. Effect of vehicle mass on measured dynamic characteristics of bridge from traffic-induced vibration test. In: Proceedings of the 19th international modal analysis conference, society for experimental mechanics, FEB 05-08, 2001. Bethel: Soc Experimental Mechanics Inc.; 2001.
11. Zhang QW, Fan LC, Yuan WC. Traffic-induced variability in dynamic properties of cable-stayed bridge. Earthquake Eng Struct Dyn. 2002;31(11):2015–21.
12. Chen J, Xu YL, Zhang RC. Modal parameters identification of using Ma suspension bridge under typhoon Victor: EMD-HT method. J Wind Eng Ind Aerodyn. 2004;92(10):805–27.
13. Ni YQ, Ko JM, Hua XG, Zhou HF. Variability of measured modal frequencies of a cable-stayed bridge under different wind conditions. Smart Struct Syst. 2007;3(3):341–56.
14. Deng Y, Ding YL, Li AQ. Quantitative evaluation of variability in modal frequencies of a suspension bridge under environmental conditions. J Vib Shock. 2011;30(8):230–6.
15. Ding YL, Deng Y, Li AQ. Study on correlations of modal frequencies and environmental factors for a suspension bridge based on improved neural networks. Sci China-Technol Sci. 2010;53(9):2501–9.
16. Hu WH, Moutinho C, Caetano E, Magalhaes F, Cunha A. Continuous dynamic monitoring of a lively footbridge for serviceability assessment and damage detection. Mech Syst Signal Process. 2012;33:38–55.
17. Yan AM, Kerschen G, De Boe P, Golinval JC. Structural damage diagnosis under varying environmental conditions—part I: a linear analysis. Mech Syst Signal Process. 2005;19(4):847–64.
18. Chen ZW, Cai QL, Lei Y, Zhu SY. Damage detection of long-span bridges using stress influence lines incorporated control charts. Sci China-Technol Sci. 2014;57(9):1689–97.

19. Kosnik DE, Zhang WZ, Durango-Cohen PL. Application of statistical process control for structural health monitoring of a historic building. J Infrastruct Syst. 2014;20(1). https://doi.org/10.1061/(asce)is.1943-555x.0000164.
20. Kullaa J. Damage detection of the Z24 Bridge using control charts. Mech Syst Signal Process. 2003;17(1):163–70.
21. Deraemaeker A, Reynders E, De Roeck G, Kullaa J. Vibration-based structural health monitoring using output-only measurements under changing environment. Mech Syst Signal Process. 2008;22(1):34–56.
22. Magalhaes F, Cunha A, Caetano E. Vibration based structural health monitoring of an arch bridge: From automated OMA to damage detection. Mech Syst Signal Process. 2012;28:212–28.
23. Wang ZR, Ong KCG. Autoregressive coefficients based Hotelling's T2 control chart for structural health monitoring. Comput Struct. 2008;86(19–20):1918–35.
24. Yi TH, Li HN, Song GB, Guo Q. Detection of shifts in GPS measurements for a long-span bridge using CUSUM chart. Int J Struct Stab Dyn. 2016;16(4):1640024.
25. Montgomery DC. Introduction to statistical quality control. New York: Wiley; 2005.
26. Quesenberry CP. On properties of binomial Q charts for variables. J Qual Technol. 1995;27(3):184–203.
27. Quesenberry CP. SPC Q charts for start-up processes and short or long runs. J Qual Technol. 1991;23(3):213–24.
28. Quesenberry CP. SPC Q charts for a binomial parameter p: short or long runs. J Qual Technol. 1991;23(3):239–46.
29. Quesenberry CP. SPC Q charts for a Poisson parameter λ: short or long runs. J Qual Technol. 1991;23(4):296–303.
30. Ren WX, Peng XL (2005) Baseline finite element modeling of a large span cable-stayed bridge through field ambient vibration tests. Comput Struct. 2005;83(8–9):536–50.
31. Ding YL, Li AQ, Sun J, Deng Y. Experimental and analytical studies on static and dynamic characteristics of steel box girder for Runyang Cable-stayed Bridge. Adv Struct Eng. 2008;11(4):425–38.
32. Macdonald JHG. Identification of the dynamic behaviour of a cable-stayed bridge from full-scale testing during and after construction. Bristol: University of Bristol; 2000.
33. Li AQ, Miao CQ, Li ZX, Han XL, Wu SD, Ji L, Yang YD. Health monitoring system for the Runyang Yangtse River Bridge. J Southeast Univ (Natural Science Edition). 2003;33(5):544–8.
34. Yourstone SA, Zimmer WJ. Non-normality and the design of control charts for averages. Decis Sci. 1992;23(5):1099–113.
35. Bowman AW, Azzalini A. Applied smoothing techniques for data analysis: the Kernel approach with S-Plus illustrations. New York: Oxford University Press; 1997.
36. Scott DW. Multivariate density estimation: theory, practice, and visualization. New York: Wiley; 1992.
37. Sohn H, Czarneck JA, Farrar CR. Structural health monitoring using statistical process control. J Struct Eng ASCE. 2000;126(11):1356–63.
38. Mackay DJC. Bayesian interpolation. Neural Comput. 1992;4(3):415–47.
39. Deng Y, Li AQ, Feng DM. Probabilistic damage detection of long-span bridges using measured modal frequencies and temperature. Int J Struct Stab Dyn. 2018;18(10):1850126.

Part 3

Chapter 5
Fatigue Monitoring of Welded Details

5.1 Introduction

The ASCE Committee on Fatigue and Fracture Reliability reported that 80–90% of failures in steel structures are related to fatigue and fracture [1, 2]. During the operational period, bridge structures are subjected to millions of cyclic loadings caused by live traffic loads, resulting in the fatigue damage and fracture taken place in the critical components of steel bridges. Welded joints in steel bridges are considered to be more vulnerable to the fatigue failure [3, 4].

Stress analysis based on numerical simulation is vastly used to evaluate the fatigue performance of steel bridges [5–10]. In general, the finite element model (FEM) of the steel bridge is first established. Then the simulated vehicle load sequences are applied directly to the FEM or the derived influence lines (surfaces) to obtain the stress histories of steel components. For example, Basso et al. [10] established the FEM of a historic railway bridge and the stress responses of the bridge subjected to train-crossing loads was simulated. Then the fatigue reliability assessment was carried out using the simulated results of stress response. In recent years, however, it is proved that structural monitoring is essential in the fatigue performance assessment of steel bridges [11, 12]. Consequently, a more accurate method to measure the fatigue stress is introduced with the development of structural health monitoring system (SHMS) [13–22]. SHMS provides opportunities to observe more realistic stress distributions and fatigue behaviors under operational excitations. For example, Li and Chan et al. [14–16] presented a fatigue damage model based on the continuum damage mechanics. By using this new model, they proposed an approach for evaluating the fatigue damage of bridge-deck sections based on 7 days' strain data from an online SHMS of Tsing Ma Bridge in Hong Kong. Xiao et al. [17] estimated the fatigue life of longitudinal ribs in steel orthotropic deck of Kinuura Bridge in Japan using 3 days' measured strain data based on linear elastic fracture mechanics.

The existing research results show that to carry out effective fatigue performance evaluation, representative blocks of stress histories should be defined and used

Y. Deng and A. Li, *Structural Health Monitoring for Suspension Bridges*,
https://doi.org/10.1007/978-981-13-3347-7_5

[14–19]. However, relatively small amount of data in the representative block may fail to cover the complexity and diversity of fatigue loading effects (including stress range and number of cycles) caused by environmental conditions [20, 22]. Therefore, long-term continuous strain monitoring is recommended for more accurate and realistic fatigue performance evaluations of steel bridges. This chapter presents a case-study focusing on the fatigue performance of welded details based on 4-year monitoring strain data since Runyang Yangtze Bridge being was open to traffic. The fatigue life prediction methodology is firstly introduced by using the Palmgren–Miner damage rule and the S–N design curve for welded details in orthotropic steel decks. Then the analysis of stress range histograms and necessary discussion of long-term monitoring are carried out. Finally, the fatigue life predictions of the welded details considering the traffic flow growth are presented.

5.2 Fatigue Stress Monitoring of the Runyang Yangtze Bridge

The Runyang Yangtze Bridge, across the Yangtze River in Jiangsu province, China, is composed of two series-lined bridges, namely, the Runyang Suspension Bridge (RSB) and the Runyang Cable-stayed Bridge (RCB). The RSB, as shown in Fig. 2.1, has been briefed in Chap. 2. The RCB, as shown in Fig. 5.1, is a three-span cable-stayed bridge, with a total length of 756.8 m. The main span is 406 m and the two side spans 175.4 m. The cross section of the deck of both RSB and RCB is aerodynamically shaped closed steel box girder, which carries two carriageways, and each has three traffic lanes. The width and height of RSB's steel box girder are 36.3 and 3.0 m, and the width and height of RCB's steel box girder are 37.4 and 3.0 m.

During the construction period, three types of strain sensors, namely, the optical fiber strain sensor, vibration chord strain sensor, and strain gauge, were installed at certain key positions in order to evaluate the fatigue performances of the welded details [20, 23]. Five critical cross sections of both the RSB and the RCB are equipped with the strain sensors. However, only the mid-span sections have been installed with strain gauges, which can be used to collect the dynamic strain data. The strain gauges in the mid-span of RSB and RCB are shown in Figs. 5.2 and 5.3, respectively. The fatigue performance of the welded details in the orthotropic decks, which are close to sensors ZLNL4-15, ZLNL2-2-23 and ZLNL2-2-24, will be given in-depth analysis in this chapter. Specifically, the strain sensors ZLNL4-15 and ZLNL2-2-23 along transverse bridge direction is used to investigate the fatigue performance of the nearby rib-to-deck details, and sensor ZLNL2-2-24 along longitudinal bridge direction for that of the nearby rib-to-rib details. It is noteworthy that the sampling frequency of the strain gauge is 20 Hz.

Fig. 5.1 Runyang Cable-stayed Bridge

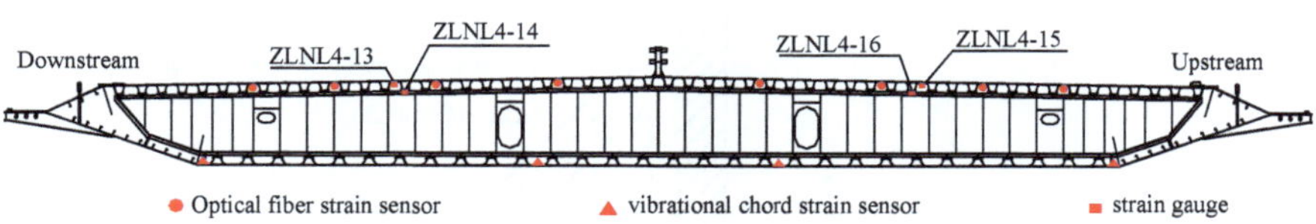

Fig. 5.2 Layout of strain sensors in the mid-span of the RSB

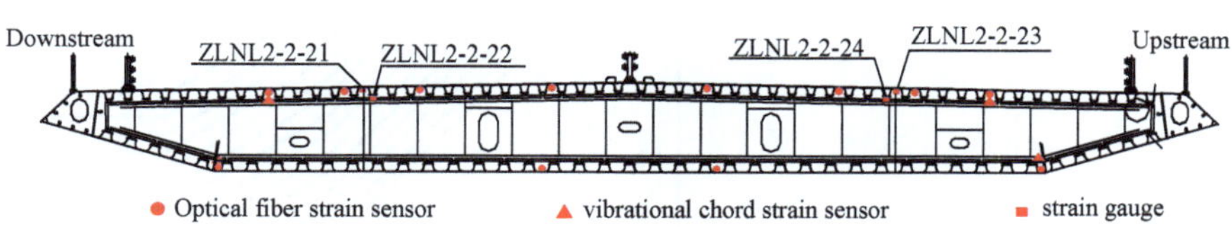

Fig. 5.3 Layout of strain sensors in the mid-span of the RCB

5.3 Fatigue Damage Determination Framework

5.3.1 S–N Curves of Welded Orthotropic Decks

Eurocode 3 provides the ordinary detail categories of the welded orthotropic decks with closed stringers and open stringers [24]. The fatigue resistance of steel box girders of the Runyang Yangtze Bridge was designed according to the S-N curves provided in Eurocode 3. Consequently, the S-N curves are also adopted in this chapter. The detail categories for the deck-to-rib weld and the rib-to-rib details considered in this work are 50 and 71, respectively, according to Table 8.8 in Eurocode 3. Thus, the corresponding S-N curves of ZLNL4-15, ZLNL2-2-23 and ZLNL2-2-24 can be determined. As shown in Fig. 5.4, the S-N curves in Eurocode 3 adopt double slopes to take into account the low-level stress cycles. When stress range is above the constant amplitude fatigue limit (CAFL), the slope of the curves is −1/3; when stress range is below the CAFL, the slope of the curves is −1/5; and when stress range is below the variable amplitude fatigue limit (VAFL), the curves become horizontal straight lines. It means that if all the stress range is below the VAFL, corresponding fatigue life of the welded details is infinite.

For nominal stress range spectrum, Eurocode 3 provides the fatigue strength curves by

$$\Delta\sigma_R^3 N_R = \Delta\sigma_C^3 \cdot 2 \times 10^6 = K_C\left(N \leq 5 \times 10^6\right) \tag{5.1a}$$

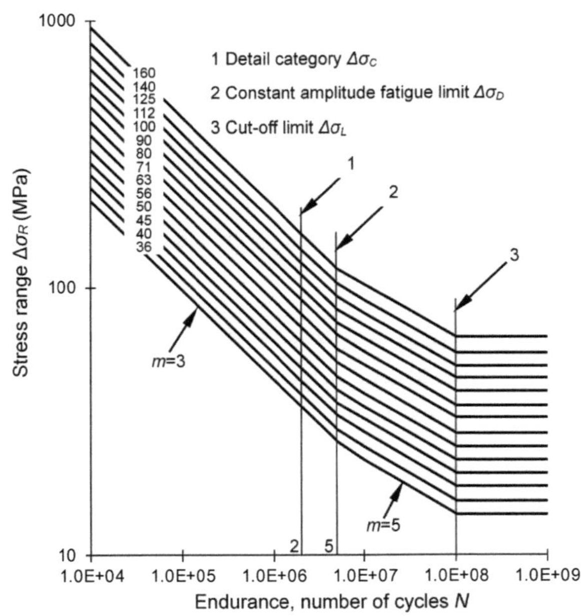

Fig. 5.4 Fatigue strength curves in Eurocode 3

$$\Delta\sigma_R^5 N_R = \Delta\sigma_D^5 \cdot 5 \times 10^6 = K_D(5 \times 10^6 \leq N \leq 10^8), \tag{5.1b}$$

where $\Delta\sigma_R$ is the stress range. N_R is the corresponding life expressed as number of cycles related to $\Delta\sigma_R$. $\Delta\sigma_C$, which is also called as detail category, is the fatigue strength when fatigue life is 2 million cycles. $\Delta\sigma_D$ is the CAFL when fatigue life is 5 million cycles. The fatigue strength coefficient is defined as K_C when $\Delta\sigma_R$ is bigger than $\Delta\sigma_D$, and as K_D when $\Delta\sigma_R$ is smaller than $\Delta\sigma_D$. According to the specifications in Eurocode 3, K_C and K_D of the deck-to-rib welded details involved in this work are 2.50×10^{11} and 3.47×10^{14}, respectively. K_C and K_D of the rib-to-rib welded details are 7.16×10^{11} and 1.90×10^{15}, respectively. In addition, Eurocode 3 presented a cutoff limit $\Delta\sigma_L$

$$\Delta\sigma_L = \left(\frac{5}{100}\right)^{\frac{1}{5}} \cdot \Delta\sigma_D = 0.549\Delta\sigma_D \tag{5.2}$$

when $\Delta\sigma_R$ is smaller than $\Delta\sigma_L$, $\Delta\sigma_R$ leads to no fatigue damage to structural component. Thus, $\Delta\sigma_L$, which is the VAFL, can be treated as the threshold of variable amplitude stress range.

5.3.2 Equivalent Stress Range and Fatigue Damage

In most cases there would exist flaws, such as microcracks, in the welds. When these flaws are exposed to low-level stress cycles, they will not develop. Once high-level stress cycles are included in the stress range spectrum, however, these flaws will develop and be further affected by low-level stress cycles. Hence, the threshold of variable amplitude stress range may not exist because of the high-level stress cycles [15]. For the highway steel box girder bridges, it is proposed that the stress range, which is below VAFL, should also be considered in the determination of fatigue life.

According to Eq. (5.1), the fatigue damage due to stress range S is

$$\frac{n}{N} = \frac{nS^3}{K_C}(\Delta\sigma_D \leq S) \tag{5.3a}$$

$$\frac{n}{N} = \frac{nS^5}{K_D}(S \leq \Delta\sigma_D), \tag{5.3b}$$

where n is the number of S in the stress range spectrum. N is the fatigue life related to stress range S. Based on the Palmgren–Miner rule [25], the fatigue damage under variable amplitude loading can be obtained. When stress range S is smaller than CAFL $\Delta\sigma_D$, the fatigue damage is

$$D = \frac{nS^5}{K_D} \tag{5.4}$$

On the other hand, when stress range S is bigger than $\Delta\sigma_D$, the fatigue damage is

$$D = \frac{nS^3}{K_C} \tag{5.5}$$

Thus, the fatigue damage of a stress range histogram can be calculated as

$$D = \sum_{\Delta\sigma_D \leq S_i} \frac{n_i S_i^3}{K_C} + \sum_{S_j \leq \Delta\sigma_D} \frac{n_j S_j^5}{K_D}, \tag{5.6}$$

where, n_i is the number of stress range S_i, which is bigger than $\Delta\sigma_D$. And n_j is the number of stress range S_j, which is smaller than $\Delta\sigma_D$. On the basis of the equivalence principle of fatigue damage, the variable amplitude stress range can be equivalent to a constant amplitude stress range, which is named as the equivalent stress range

$$D = \frac{N_d \cdot S_{\text{eq}}^5}{K_D} \tag{5.7}$$

The expressions of equivalent stress range S_{eq} and corresponding number of cycle N_d are

$$S_{\text{eq}} = \left(\frac{\sum_{\Delta\sigma_D \leq S_i} \frac{n_i S_i^3}{K_C} + \sum_{S_j \leq \Delta\sigma_D} \frac{n_j S_j^5}{K_D}}{N_d / K_D} \right)^{1/5} \tag{5.8}$$

$$N_d = \sum_{\Delta\sigma_D \leq S_i} n_i + \sum_{S_j \leq \Delta\sigma_D} n_j \tag{5.9}$$

5.4 Processing of Strain Measurements

5.4.1 Original Strain Data Analysis

The strain data from 2006 through 2009, recorded by the three above-mentioned sensors, are utilized to investigate the fatigue performance of the welded details. To simplify the expression, ZLNL2-2-23 denotes not only the specific strain gauge in the SHMS but also the nearby rib-to-deck welded detail in the steel box girder of RCB, and so are ZLNL4-15 and ZLNL2-2-24.

One-day data of ZLNL2-2-23 is first used to analyze the characteristics of the original strain monitoring data. The strain history with 1,728,000 data points, recorded on August 13, 2006, is shown in Fig. 5.5. In addition, in order to investigate the components in original strain data more clearly, one-minute (from 00:00 to 00:01) segment of the strain history is enlarged, as shown in Fig. 5.6.

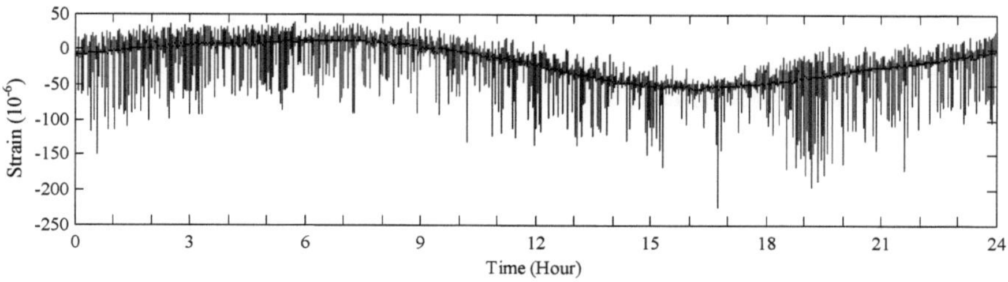

Fig. 5.5 Original strain history measured by sensor ZLNL2-2-23 on August 13, 2006

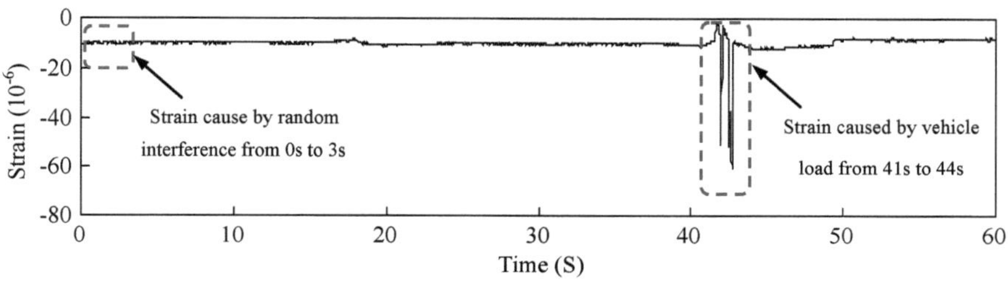

Fig. 5.6 Strain history of sensor ZLNL2-2-23 on August 13, 2006 from 00:00 to 00:01

As shown in Figs. 5.5 and 5.6, the original strain data are originated from three excitation sources, namely, from temperature fluctuation, environmental random interference, and traffic loads, specifically

(i) The first component is the diurnal variation of the mean strain induced by temperature fluctuation. Figure 5.7 gives a daily temperature history from a temperature sensor on August 13, 2006. Figure 5.7 also plots the mean strain history of sensor ZLNL2-2-23 extracted from the original strain data in Fig. 5.5 using the wavelet transform technique [26]. From 6:00 to 7:00 the temperature of steel box girder is relatively low, but the mean strain is large. Furthermore, from 14:00 to 15:00 the temperature of steel box girder is relatively high, but the mean strain is small. In addition, there is a small-time delay of the maximum or minimum mean strain when compared with the appearance of the maximum or minimum temperature. Figure 5.7 reveals that the main strain of the welded detail seems to be influenced strongly by structural temperature.

(ii) The second component is the random interference which is probably caused by operational environment or monitoring process. Figure 5.8 plots an enlarged segment of the strain history in Fig. 5.6, which represents a typical strain component mainly caused by the random interference. In Fig. 5.8, there exist quite a bit peaks and valleys in the strain curve, varying between $-10.6\mu\varepsilon$ and $-9.8\mu\varepsilon$.

(iii) The third component is caused by the moving vehicle loads. This strain component is definitely the most important source for fatigue damage. In most cases the vehicle loads are difficult, if not impossible, to be clearly identified using

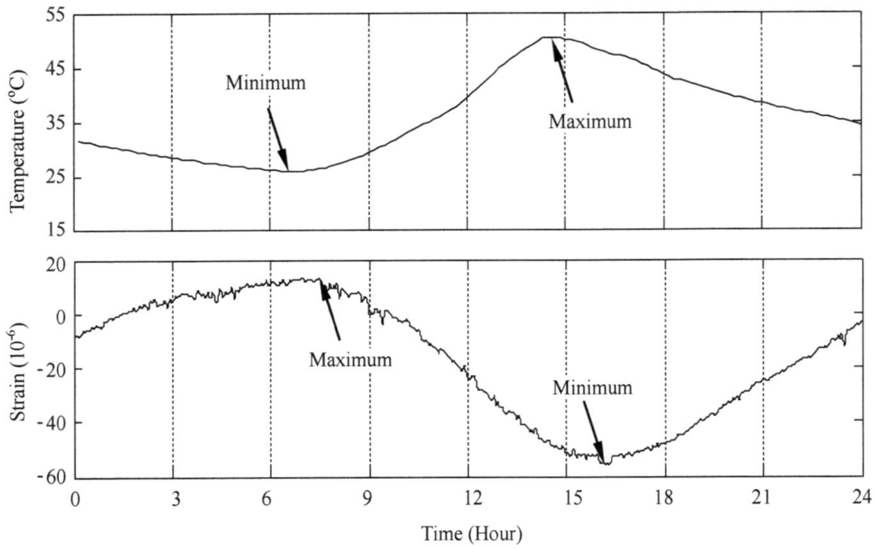

Fig. 5.7 Variation of temperature of steel box girder and mean strain of ZLNL2-2-23 on August 13, 2006

Fig. 5.8 Strain history from 0 to 3 s on August 13, 2006

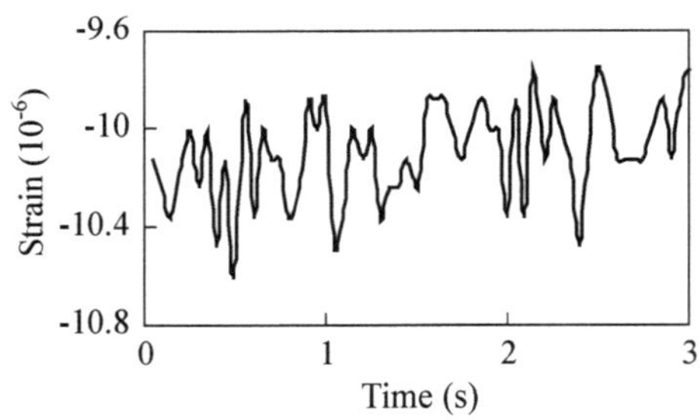

the measured strain data. Here, one-minute strain data measured in the early hours of the morning on August 13, 2006 is used to analyze the strain component caused by vehicle loads. Figure 5.6 reveals that due to an extreme small traffic flow there was only one vehicle passing through the bridge during the time from 00:00 to 00:01. Figure 5.9 shows an enlarged segment of the strain history, with four outstanding valleys, which is induced by a four-axle vehicle from 41 s to 44 s.

By using the rain-flow counting method [27], the stress ranges from the strain histories in Figs. 5.8 and 5.9 can be obtained, as shown in Fig. 5.10. It is noted that the elastic modulus of 200,000 N/mm^2 is used. For the data in Fig. 5.8, the number of cycles is 18 and all stress ranges are less than 2 MPa. It is obvious that these strain ranges from 0 to 3 s are resulted from random interference. However, the rain-flow counting results of strain history from 41 to 44 s are a bit different, with a

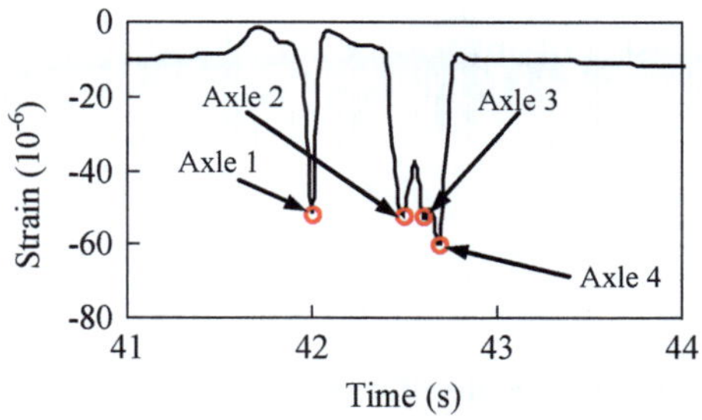

Fig. 5.9 Strain history from 41 to 44 s on August 13, 2006

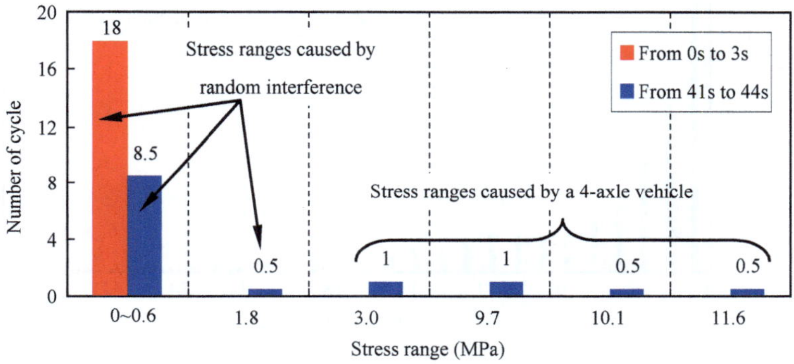

Fig. 5.10 Stress range histogram

total number of 12 cycles including 9 cycles below 2 MPa. Hence, all the rest of the stress ranges, which are larger than 2 MPa, would be caused by the four-axle vehicle. Based on the above analysis, the stress cycles caused by random interference have a large number but small range values, while the stress cycles caused by vehicle loads are the very opposite. Therefore, the enormous strain cycles with small stress range, which have no contribution to the fatigue damage, may lead to false number of cycle in calculation.

5.4.2 Temperature Effect on Stress Range Histogram

Figure 5.11 plots the strain history after removing the effect of temperature using the wavelet transform decomposition technique. Then the strain histories in both Figs. 5.5 and 5.11 can be transformed into the stress histories by multiplying the elastic modulus. Figure 5.12 shows the stress range spectrums of original data and the data without temperature effect. It can be seen that the two stress range spectrums are almost identical. Hence, the influence of temperature variation on the stress range

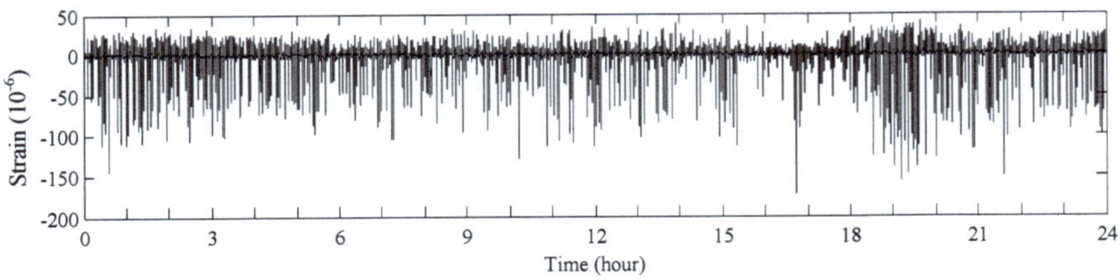

Fig. 5.11 Strain history after removing the effect of temperature (ZLNL2-2-23 on August 13, 2006)

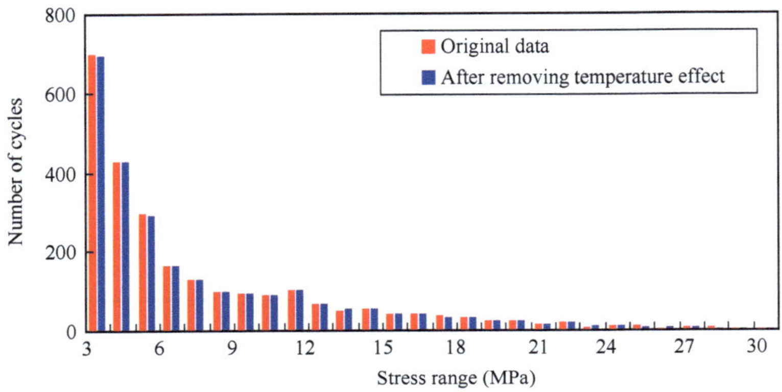

Fig. 5.12 Stress range histograms of sensor ZLNL2-2-23 on August 13, 2006

spectrum can be ignored. Now the key problem is how to eliminate the enormous small strain points, which are caused by the random interference, from the original data.

5.4.3 *Random Interference in Stress Range Histogram*

Figure 5.13 shows the stress range histogram of ZLNL2-2-23 on August 13, 2006. This figure includes an enlarged histogram from 2 to 40 MPa. Figure 5.14 gives the corresponding fatigue damage histogram calculated from Eqs. (5.4) or (5.5) depending on the values of its stress ranges. There are a considerable number of low-level stress cycles in the stress range histogram. However, the fatigue damage caused by these low-level stress cycles is relatively small and most of the fatigue damage is resulted from the high-level stress cycles. For example, the number of the stress ranges from 0 to 2 MPa is 469,998, only resulting in a total fatigue damage of 1.6e−9, while the number of the stress ranges from 9 to 10 MPa is 91 with a much larger fatigue damage of 2.6e−8. These low-level stress cycles would result in erroneous equivalent stress range and number of cycles.

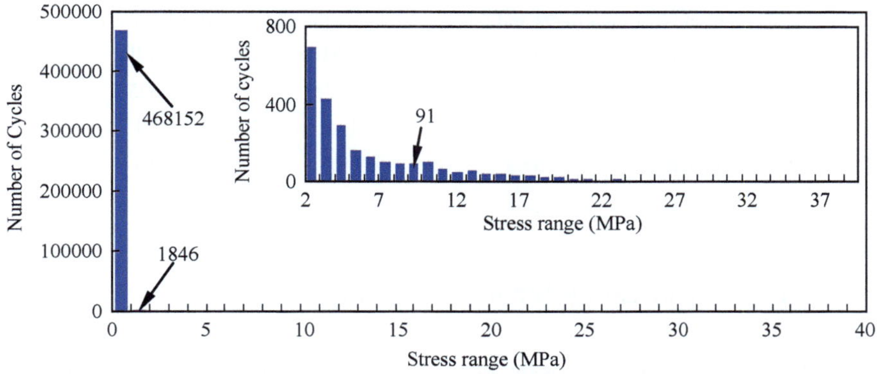

Fig. 5.13 Stress range histogram of sensor ZLNL2-2-23 on August 13, 2006

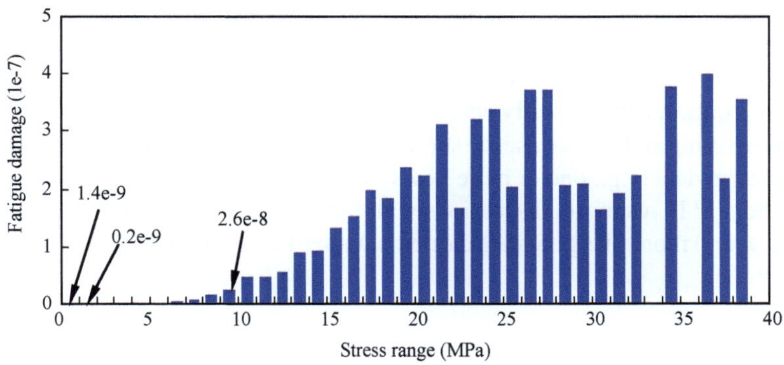

Fig. 5.14 Fatigue damage histogram of sensor ZLNL2-2-23 on August 13, 2006

Hence, it is necessary to define a threshold to eliminate the random interference from measured stress range histograms. The stress ranges, which are less than the threshold, will be omitted during calculation of the number of cycles, equivalent stress range and cumulative fatigue damage. When threshold is 0 MPa, it means all the stress ranges involved. Table 5.1 lists number of cycle, equivalent stress range and cumulative fatigue damage with three different thresholds. Based on the results in Figs. 5.13, 5.14 and Table 5.1, it is found that the vast majority of the stress ranges from 0 to 2 MPa caused by random interference contribute little to the cumulative fatigue damage. Therefore, the threshold of stress range can be set to be 2 MPa so as to eliminate its influence to obtain more reasonable equivalent stress range and number of cycles.

Table 5.1 The fatigue loading effects and fatigue damage with 3 thresholds of stress range

Threshold of stress range	0 MPa	1 MPa	2 MPa
Number of cycle	472,623	4471	2625
Equivalent stress range	5.4	13.6	15.1
Cumulative fatigue damage (1e−7)	59.9	59.8	59.8

5.5 Necessity of Long-Term Monitoring

Monitoring strain data provides the possibility to predict when, where and how the fatigue failure might occur. Generally, in order to carry out long-term, continual and real-time monitoring and evaluation, the SHMS for long-span bridges need a large number of instrumentations, including sensing devices, collecting units, transmission equipment, and controlling software. Most importantly, the mass data collected and stored by the monitoring systems, need to be well analyzed, which further raises the following questions: is it really necessary to establish a costly system for the long-term monitoring and can we use the short-term data (one hour or one day, etc.) instead of long-term data (e.g., one or more years) to investigate the fatigue performance of steel bridges? In response, discussions with the medium-term (3 months) and long-term (3 years) monitoring data are carried out after the studies on the short-term (one day) data.

5.5.1 Medium-Term Monitoring

The data of ZLNL2-2-23 from July to September of 2007 is adopted here. As shown in Fig. 5.15, the Jiangyin Suspension Bridge, also crossing the Yangtze River in China, was closed for maintaining during August 2007. The Jiangyin Suspension bridge, with a main span of 1385 m, connects the cities of Jiangyin and Jinjiang in Jiangsu province. The locations of the Jiangyin Suspension Bridge and Runyang Yangtze Bridge are shown in Fig. 5.16.

Figure 5.17 shows the daily number of cycles, daily equivalent stress range and daily fatigue damage of ZLNL2-2-23 in these three months. Accordingly, the monthly averaged values are also given in Table 5.2. The results show that there was a substantial increase of the daily number of cycles in August over July and September. This increase can be explained by the significant increase of the traffic flow of Runyang Yangtze Bridge when Jiangyin Bridge was closed. In addition, the daily equivalent stress range also experienced considerable changes during these three months. The monthly averaged equivalent stress range in August increased by 8 and 13% respectively compared with that in July and in September. Moreover, the results show that the daily fatigue damage increased significantly in August and there were 98 and 181% growths in August over July and September, respectively.

Fig. 5.15 Jiangyin Suspension Bridge

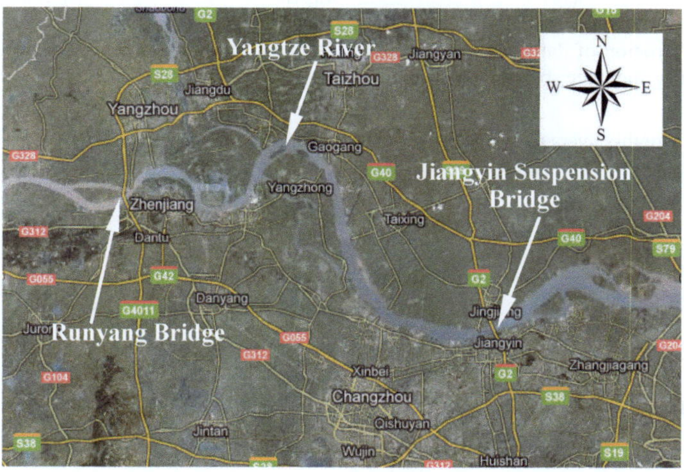

Fig. 5.16 Bridge locations

The above analysis indicates that the fatigue damage of highway steel box girder bridges can be readily affected by the temporary changes of traffic conditions. Therefore, the medium-term data collected during these 3 special months are not appropriate to be used for the fatigue life prediction.

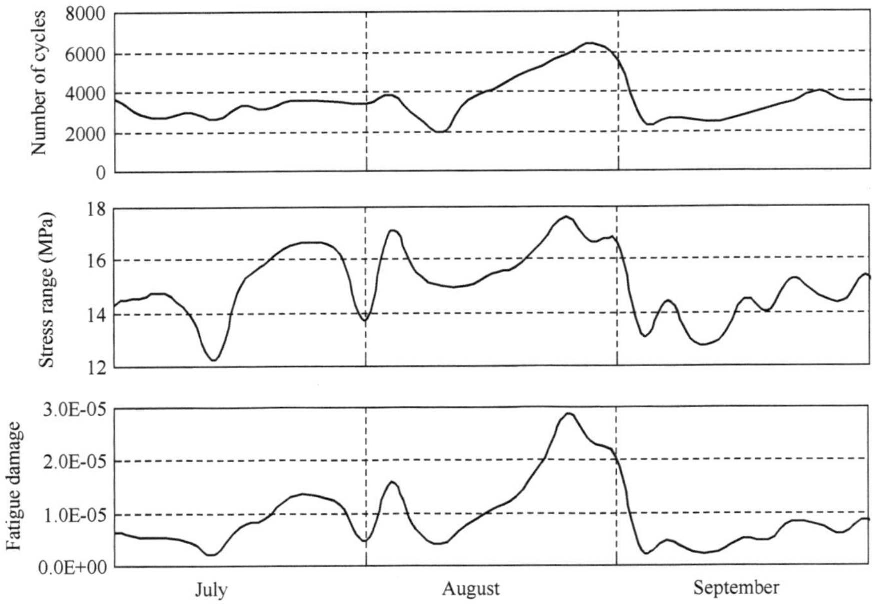

Fig. 5.17 Variation of daily number of cycles, daily equivalent stress range, and daily fatigue damage during the three special months in 2007

Table 5.2 Monthly averaged value of daily fatigue loading effects and damage

Time interval	Number of cycles	Equivalent stress range (MPa)	Fatigue damage (1e−06)
July, 2007	3182	14.9	7.60
August, 2007	4394	16.1	15.07
September, 2007	3048	14.2	5.36

5.5.2 Long-Term Monitoring

Herein, 3-year data of ZLNL2-2-23 in 2007 and 2009 is employed to further prove the necessity of long-term monitoring. Figures 5.18 and 5.19 show the annual variation curves of daily equivalent stress range and daily fatigue damage in 2009, respectively. These figures show that both equivalent stress range and fatigue damage vary greatly in a regular pattern. The equivalent stress range and fatigue damage during the summer are the largest, with the spring and autumn the second, and winter the third. This pattern can result in great influence on the prediction of fatigue life. Based on different segment choices of the monitored data, Table 5.3 gives the fatigue life prediction results of ZLNL2-2-23. For example, if the strain data from June to July of 2009 are used, the average daily fatigue damage is the largest, corresponding to the smallest predicted fatigue life of 337 years. If 1-year strain data from January

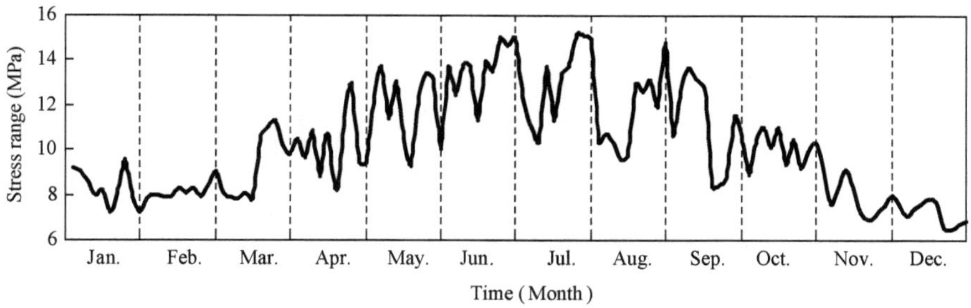

Fig. 5.18 Daily equivalent stress range of ZLNL2-2-23 in 2009

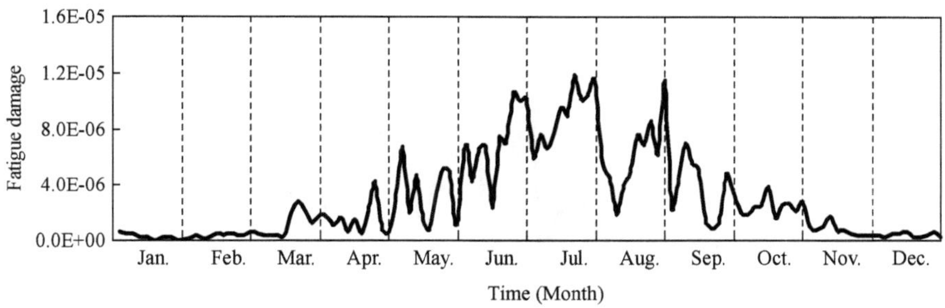

Fig. 5.19 Daily fatigue damage of ZLNL2-2-23 in 2009

Table 5.3 The effects of data length on fatigue life prediction of ZLNL2-2-23 in 2009

Length of data	Jun. to Jul.	May. to Aug.	Apr. to Sep.	Mar. to Oct.	Feb. to Nov.	Jan. to Dec.
Averaged daily fatigue damage (1e−06)	8.14	6.43	5.13	4.30	3.51	3.01
Prediction of fatigue life (year)	337	426	534	637	780	910

to December is chosen, the predicted fatigue life is 910 years, which is almost 2.7 times as big as the result of data from June to July.

In addition, Figs. 5.20 and 5.21 show the daily fatigue damage of ZLNL2-2-23 in the years of 2007 and 2008. It can be seen that the fatigue damage in these 2 years changed in the similar pattern as that in 2009. The daily fatigue damage in 2007 had a notable increase in August because the Jiangyin Suspension Bridge was closed in this month for maintaining. The average daily fatigue damage in the years of 2007, 2008, and 2009 are 3.27e−06, 2.95e−06, and 3.01e−06, respectively. The results

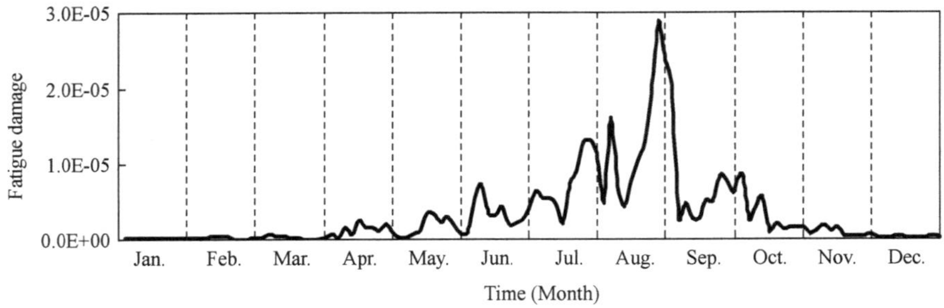

Fig. 5.20 Daily fatigue damage of ZLNL2-2-23 in 2007

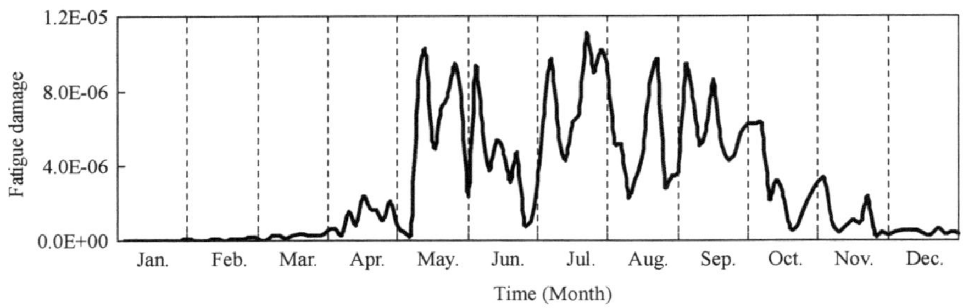

Fig. 5.21 Daily fatigue damage of ZLNL2-2-23 in 2008

show that the daily fatigue damage, which was averaged from 1-year data, changed a little.

The above analysis on the necessity of long-term monitoring reveals that there are remarkable variations in daily fatigue damage of steel highway bridges. Short-term data monitored during a few days or medium-term data during a few months are not adequate enough to demonstrate the actual fatigue behaviors of steel bridges. The prediction results may be far from the actual fatigue life if the short-term or medium-term monitoring data is employed. Hence, it is necessary to carry out the long-term monitoring (at least 1 year) for the steel highway bridge and thus the large amounts of data will be accumulated to obtain more accurate predicted fatigue life.

5.6 Fatigue Life Prediction

In developing countries like China, the traffic volume would be continuously increasing during a rather long period. Thus, the fatigue life prediction needs to account for the traffic flow growth. In order to investigate the impact of increased loads on the fatigue detail's reliability, Righiniotis [28] presented several assumptions regarding the increase in the stress range, number of cycles, and maximum loads. One of the traffic growth patterns is to assume that over a certain period, growth takes place in a

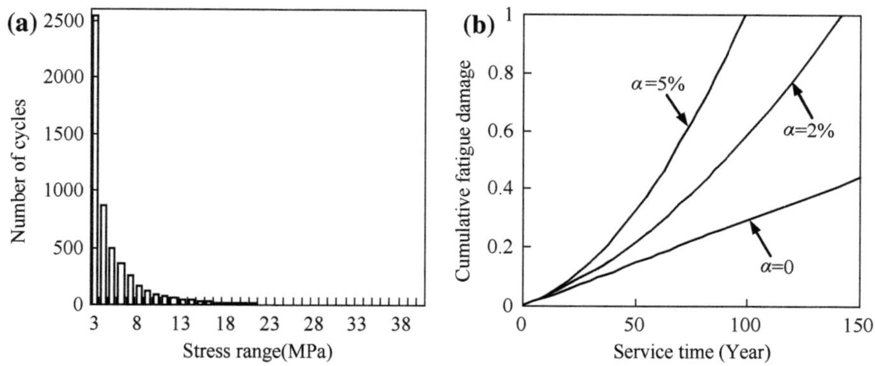

Fig. 5.22 Fatigue life evaluation results of welded detail ZLNL4-15: **a** averaged stress range histogram and **b** development of fatigue damage

linear fashion. Deng et al. [20] directly adopted the linear growing pattern of number of cycles to consider the traffic growth. In this work the traffic flow growth is also simulated with a linear increase of number of cycles. The daily number of cycles in the year of n is defined as

$$N_n = N_d + (n - 1) \cdot \alpha \cdot N_d, \tag{5.10}$$

where α is coefficient of the linear traffic flow growth. The traffic flow growth coefficients are assumed to be 0, 2 and 5%, respectively. N_d is the original daily number of cycles. Then the cumulative number of cycles during the service time is [29]

$$N_{\text{total}} = 365 \cdot N_d \cdot n \cdot \left[1 + \frac{(n-1) \cdot \alpha}{2}\right] \tag{5.11}$$

According to Eq. (5.7), the cumulative fatigue damage D_{total} is

$$D_{\text{total}} = \frac{N_{\text{total}} \cdot S_{\text{eq}}^5}{K_D} = 365 \cdot N_d \cdot n \cdot \left[1 + \frac{(n-1) \cdot \alpha}{2}\right] \cdot \frac{S_{\text{eq}}^5}{K_D}, \tag{5.12}$$

when D_{total} equals to 1, n is the fatigue life.

Figures 5.22a, 5.23a and 5.24a show the averaged stress range histograms of ZLNL4-15, ZLNL2-2-23, and ZLNL2-2-24 based on the strain data collected during 4 years from 2006 to 2009. In addition, Figs. 5.22b, 5.23b and 5.24b give the fatigue damage development of the three welded details with traffic flow growth. The figures indicate that fatigue damage develops linearly and slowly when traffic flow growth is not concerned. In this situation, the fatigue damages of the three details are all less than 1 when the service time equals to the design reference period (100 years). When traffic flow growth is concerned, fatigue damage increases nonlinearly and rapidly.

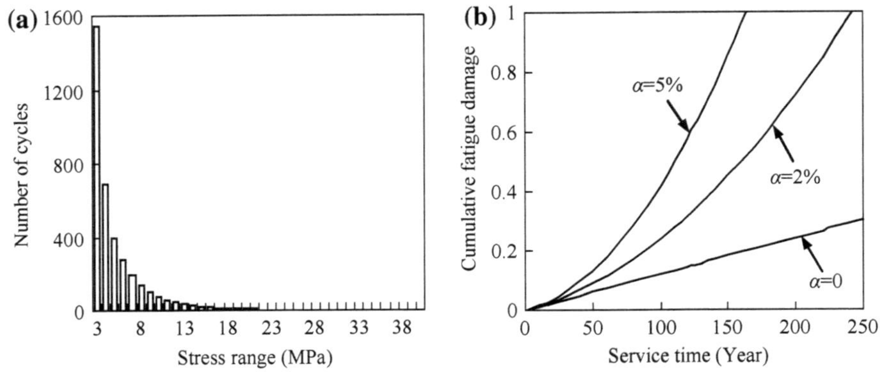

Fig. 5.23 Fatigue life prediction results of welded detail ZLNL2-2-23: **a** averaged stress range histogram and **b** development of fatigue damage

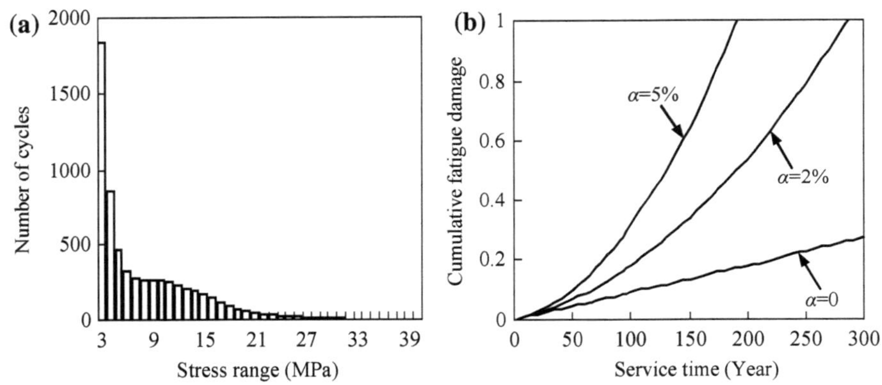

Fig. 5.24 Fatigue life prediction results of welded detail ZLNL2-2-24: **a** averaged stress range histogram and **b** development of fatigue damage

Table 5.4 summarizes the fatigue evaluation results of the three details. The results show that the fatigue service time can basically meet the requirement of the design reference period even if the traffic flow growth is concerned. The fatigue service time of ZLNL4-15, the value of which is 99 years, is a little less than the design reference period only when the traffic flow growth coefficient is 5%. For the same kind of welded detail, the fatigue damage of ZLNL4-15 in RSB develops more quickly than that of ZLNL2-2-23 in RCB. For the two details in RCB, the fatigue life of ZLNL2-2-23 is shorter than that of ZLNL2-2-24.

Table 5.4 Fatigue evaluation results of the welded details

Welded type	Welded details	Fatigue damage when service time is 100 years			Fatigue life (year) when fatigue damage is 1		
		$\alpha=0$	$\alpha=2\%$	$\alpha=5\%$	$\alpha=0$	$\alpha=2\%$	$\alpha=5\%$
Rib-to-deck	ZLNL4-15	0.293	0.584	>1	341	142	99
	ZLNL2-2-23	0.121	0.240	0.419	829	245	165
Rib-to-rib	ZLNL2-2-24	0.091	0.180	0.314	1107	285	195

5.7 Summary

This chapter presents a case-study of the fatigue performance investigation of welded details in large-span steel bridges utilizing the long-term monitoring strain data collected by the SHMS. The following conclusions can be drawn from the study

1. The strain data is originated from three excitation sources, namely, from temperature fluctuation, environmental random interference, and traffic loads. The analysis results reveal that the effect of temperature on the stress range spectrum is very small. This effect can be ignored during fatigue evaluation. In the stress range histogram, a large number of small strain ranges mainly caused by random interference, contribute little to the fatigue damage of welded details. This study proposes a threshold of 2 MPa for the stress range to eliminate the influence of random interference.
2. During the closing period of the Jiangyin Suspension Bridge, the fatigue damage of ZLNL2-2-23 is twice larger than that during normal operational period. Moreover, for ZLNL2-2-23 the predicted fatigue life using the 1-year data of 2009 is 2.7 times bigger than that using the data from June to July in 2009. The further analysis shows that the daily fatigue damage, which was averaged from 1-year data, changed a little. Therefore, short-term- and medium-term monitoring data cannot provide enough information to present an accurate fatigue life prediction. The results demonstrate that it is necessary to conduct the long-term strain monitoring for welded details of steel box girder bridges.
3. The fatigue life prediction with consideration of traffic flow growth is presented. The results show that traffic flow growth can significantly accelerate the fatigue damage development rate. Hence, the monitoring systems should pay attention to the change of the traffic volume during the service time.

References

1. ASCE Committee on fatigue and fracture reliability of the committee on structural safety and reliability of the structural division. Fatigue reliability 1–4. J Struct Div ASCE. 1982;108(1):3–88.
2. Byers WG, Marley MJ, Mohammadi J, Nielsen RJ, Sarkani S. Fatigue reliability reassessment applications: state-of-the-art paper. J Struct Eng ASCE. 1997;123(3):277–85.
3. Tsakopoulos PA, Fisher JW. Full-scale fatigue tests of steel orthotropic decks for the Williamsburg Bridge. J Bridge Eng. 2003;8(5):323–33.
4. Connor RJ, Fisher JW. Consistent approach to calculating stresses for fatigue design of welded rib-to-web connections in steel orthotropic bridge decks. J Bridge Eng. 2006;11(5):517–25.
5. MacDougall C, Green MF, Shillinglaw S. Fatigue damage of steel bridges due to dynamic vehicle loads. J Bridge Eng. 2006;11(3):320–8.
6. Mori T, Lee HH, Kyung KS. Fatigue life estimation parameter for short and medium span steel highway girder bridges. Eng Struct. 2007;29(10):2762–74.
7. Xiao ZG, Yamada K, Ya S, Zhao XL. Stress analyses and fatigue evaluation of rib-to-deck joints in steel orthotropic decks. Int J Fatigue. 2008;30(8):1387–97.
8. Pipinato A, Pellegrino C, Modena C. Fatigue assessment of highway steel bridges in presence of seismic loading. Eng Struct. 2011;33(1):202–9.
9. Guo T, Frangopol DM, Chen Y. Fatigue reliability assessment of steel bridge details integrating weigh-in-motion data and probabilistic finite element analysis. Comput Struct. 2012;112–113(4):245–57.
10. Basso P, Casciati S, Faravelli L. Fatigue reliability assessment of a historic railway bridge designed by Gustave Eiffel. Struct Infrastruct Eng. 2015;11(1):27–37.
11. DeWolf JT, Lauzon RG, Culmo MP. Monitoring bridge performance. Struct Health Monit Int J. 2002;1(2):129–38.
12. Ko JM, Ni YQ. Technology development in structural health monitoring of large-scale bridge. Eng Struct. 2005;27(12):1715–25.
13. Sartor RR, Culmo MP, DeWolf JT. Short-term strain monitoring of bridge structures. J Bridge Eng. 1999;4(3):157–64.
14. Li ZX, Chan THT, Ko JM. Fatigue analysis and life prediction of bridges with structural health monitoring data—Part I: methodology and strategy. Int J Fatigue. 2001;23(1):45–53.
15. Chan THT, Li ZX, Ko JM. Fatigue analysis and life prediction of bridges with structural health monitoring data—Part II: applications. Int J Fatigue. 2001;23(1):55–63.
16. Li ZX, Chan THT, Zheng R. Statistical analysis of online strain response and its application in fatigue assessment of a long-span steel bridge. Eng Struct. 2003;25(14):1731–41.
17. Xiao ZG, Yamada K, Inoue J, Yamaguchi K. Fatigue cracks in longitudinal ribs of steel orthotropic deck. Int J Fatigue. 2006;28(4):409–16.
18. Ni YQ, Ye XW, Ko JM. Monitoring-based fatigue reliability assessment of steel bridges: analytical model and application. J Struct Eng. 2010;136(12):1563–73.
19. Ye XW, Ni YQ, Wong KY, Ko JM. Statistical analysis of stress spectra for fatigue life assessment of steel bridges with structural health monitoring data. Eng Struct. 2012;45:166–76.
20. Deng Y, Ding YL, Li AQ, Zhou GD. Fatigue reliability assessment for bridge welded details using long-term monitoring data. Sci China Technol Sci. 2011;54(12):3371–81.
21. Li S, Zhu S, Xu YL, Chen ZW, Li H. Long-term condition assessment of suspenders under traffic loads based on structural monitoring system: application to the Tsing Ma Bridge. Struct Control Health Monit. 2012;19(1):82–101.
22. Li AQ, Ding YL, Wand H, Guo T. Analysis and assessment of bridge health monitoring mass data—progress in research/development of "Structural Health Monitoring". Sci China Technol Sci. 2012;55(8):2212–24.
23. Li AQ, Miao CQ, Li ZX, Han XL, Wu SD, Ji L, Yang YD. Health monitoring system for the Runyang Yangtse River Bridge. J Southeast Univ (Natural Science Edition). 2003;23(5):544–8.

24. British Standards Institution. Eurocode 3: design of steel structures—Part 1–9. Fatigue. 2005.
25. Miner MA. Cumulative damage in fatigue. J Appl Mech T ASME. 1945;12:159–64.
26. Reda Taha MM, Noureldin A, Lucero JL, Baca TJ. Wavelet transform for structural health monitoring: a compendium of uses and features. Struct Health Monit Int J. 2006;5(3):267–95.
27. Nieslony A. Determination of fragments of multiaxial service loading strongly influencing the fatigue of machine components. Mech Syst Signal Process. 2009;23(8):2712–21.
28. Righiniotis TD. Effects of increasing traffic loads on the fatigue reliability of a typical welded bridge detail. Int J Fatigue. 2006;28(8):873–80.
29. Deng Y, Liu Y, Feng DM, Li AQ. Investigation of fatigue performance of welded details in long-span steel bridges using long-term monitoring strain data. Struct Control Health Monit. 2015;22(11):1343–58.

Chapter 6
Fatigue Reliability Analysis for Welded Details

6.1 Introduction

Orthotropic steel decks (OSDs) have been widely adopted for long-span bridges due to its notable advantages, such as light weight, high strength and durability, and rapid construction [1, 2]. However, various types of cracking in the OSDs have been reported owing to lack of knowledge in its fatigue characteristics, design defects, and harsh loading conditions such as heavy-duty vehicles and high-density traffic volumes [1, 3, 4]. In the past few decades, significant efforts have been made on the design of S-N curves and development of fatigue life prediction approaches [5–7].

Fatigue reliability evaluation for the in-service OSD bridges requires accurate measurement of fatigue stress spectra, which can be obtained using the structural health monitoring system (SHMS) [8–10]. Existing SHMS-based fatigue evaluation approaches for the OSDs can be classified into two categories. The first type is to develop fatigue stress spectra using the finite element (FE) analysis based on measured operational vehicle flows by weigh-in-motion systems [10, 11]. Because of complex structural behaviors and randomness in the actual traffic loadings, this type method relies on the accuracy of reconstructed vehicle loading models and developed FE models [10]. The second type is directly based on the strain measurements collected by strain sensors [8, 9]. For example, Guo et al. [8] investigated effects of ambient temperature and traffic volume on fatigue damages of welded steel deck details using long-term strain monitoring data. De Freitas et al. [9] applied strain measurements from controlled load tests and 1-year field monitoring to investigate the fatigue performance of an OSD strengthened by bonding a second steel plate to the existing deck. Above strain monitoring-based studies mainly focused on deterministic analyses.

In this chapter, to consider uncertainties in fatigue analysis, a reliability-based fatigue evaluation approach for welded details of orthotropic steel decks is proposed using both the fatigue strength curves from existing specification and the long-term monitoring strain data.

Y. Deng and A. Li, *Structural Health Monitoring for Suspension Bridges*,
https://doi.org/10.1007/978-981-13-3347-7_6

6.2 Framework of the Fatigue Reliability Analysis

6.2.1 Fatigue Limit State Function

Eurocode 3 provides the fatigue detail categories of welded orthotropic decks with either closed ribs or open ribs [5]. The S-N curves in Eurocode 3 adopt double slopes to consider the low-level stress cycles. As shown in Fig. 5.4, when stress range is above the constant amplitude fatigue limit (CAFL) $\Delta\sigma_D$, the slope of the curves is $-1/3$; when stress range is below the CAFL, the slope of the curves is $-1/5$; and when stress range is below the variable amplitude fatigue limit (VAFL) $\Delta\sigma_L$, the curves become horizontal straight lines. $\Delta\sigma_C$ is the detail category. Two types of weld details for OSDs, referring to rib-to-deck and the rib-to-rib details, respectively, will be investigated in this study. The S-N curve parameters for these welds are listed in Table 6.1.

Section 5.3 have proposed a fatigue damage determination framework for welded details of OSDs. On the basis of the equivalence principle of fatigue damage, the variable amplitude stress range can be equivalent to a constant amplitude stress range, which is named as the equivalent stress range. The expression of fatigue damage is presented as Eq. (5.7).

For the fatigue reliability analysis by using monitoring data, the limit state function can be established on the basis of Eq. (5.7)

$$G(X) = \Delta - e \cdot D = \Delta - e \cdot \frac{N_c S_{\text{eq}}^5}{K_D} = 0 \tag{6.1}$$

where Δ, the critical damage, represents the fatigue resistance, e is an error coefficient for field measurements [12, 13]. The accumulated number of stress cycles N_c is often treated as a deterministic variable and in Y years it can be estimated by

$$N_c(Y) = 365 \cdot Y \cdot \text{ADSC} \tag{6.2}$$

where ADSC is the average daily stress cycles, which can be determined using monitoring data. The parameters in Eq. (6.1) can be treated as random variables to account for the uncertainties in both monitoring data and S-N curves. From existing literatures [12–15], the lognormal distribution can be used to quantify the uncertainties in the error coefficient e [12, 13], the critical damage Δ [14], and the fatigue strength coefficient K_D [15]. The statistics of these variables are summarized in Table 6.2.

Table 6.1 S-N curves of Eurocode 3

Description of welds	Detail category $\Delta\sigma_C$ (MPa)	CAFL $\Delta\sigma_D$ (MPa)	Cut-off limit $\Delta\sigma_L$ (MPa)	Fatigue strength coefficient K_D
Rib-to-deck	50	37	20	3.47×10^{14}
Rib-to-rib	71	52	29	1.90×10^{15}

Table 6.2 Statistic information of parameters

Random variable	Description	Distribution type	Mean value	COV	Source
K_D	Rib-to-deck	Lognormal	3.47×10^{14}	0.45	Zhao et al. [15], Eurocode 3 [5]
	Rib-to-rib	Lognormal	1.90×10^{15}	0.45	
Δ	Critical damage	Lognormal	1.0	0.3	Wirsching [14]
e	Measurement error coefficient	Lognormal	1.0	0.03	Frangopol et al. [12, 13]
N_c	Accumulated number of stress cycles	Deterministic	–	–	SHM data

6.2.2 Probabilistic Model for the Equivalent Stress Range

Most studies indicated that the distribution of the S_{eq} from monitoring data by SHMS is unimodal [16, 17]. However, the authors' recent studies based on field measurements have demonstrated that the S_{eq} for OSD fatigue details is distributed multimodally [18]. Hence, both the lognormal model and Gaussian mixture model (GMM) are employed to account for the actual distribution characteristics of the measured equivalent stress ranges.

GMM is one of the finite mixture distributions (FMD). The FMD has been used to quantify the uncertainty of the measured stress ranges in fatigue analysis of steel bridges [19, 20]. For example, Ni et al. [19] fitted the FMDs to the measured stress range histograms of the Tsing Ma Bridge, in which several peaks are available which can be attributed to the highway traffic, the railway traffic, and the wind excitations [20]. In this study, the GMM is used to quantify the uncertainty of the measured daily equivalent stress ranges rather than the measured stress range histograms.

The probability density distribution of the GMM, which contains M Gaussian components, can be expressed as [21]

$$f(x|\boldsymbol{\theta}) = \sum_{i=1}^{M} w_i N\left(x|\mu_i, \sigma_i^2\right) = \sum_{i=1}^{M} w_i \frac{1}{\sqrt{2\pi}\sigma_i} \cdot \exp\left[-\frac{(x-\mu_i)^2}{2\sigma_i^2}\right] \tag{6.3}$$

$$\sum_{i=1}^{M} w_i = 1 \tag{6.4}$$

where $N\left(x|\mu_i, \sigma_i^2\right)$ is the ith Gaussian component, of which the mean value and the variance are μ_i and σ_i^2, respectively. w_i denotes the weight of the ith component. The unknown parameter vector $\boldsymbol{\theta} = \left(w_1, w_2, \ldots, w_M; \mu_1, \mu_1, \ldots, \mu_1; \sigma_1^2, \sigma_2^2, \ldots, \sigma_M^2\right)$ is usually estimated by using the expectation maximization (EM) algorithm [22].

The Gaussian component number M should be priori information before estimating the unknown parameter vector $\boldsymbol{\theta}$ with the EM algorithm. Although more Gaussian components can enable more accurate fitting results, subjective selections of the number of Gaussian component are inappropriate. Obviously, inadequate Gaussian components lead to inaccurate fitting results. On the other hand, excessive Gaussian components result in unnecessary complexity in the probability models. Hence, the Akaike information criterion (AIC) [23] and the Bayesian information criteria (BIC) [24] are employed to search the optimal Gaussian component number to achieve both accuracy and conciseness. The expressions of the AIC and BIC can be given as

$$\mathrm{AIC} = 2m - \ln L(x|M, \boldsymbol{\theta}) \tag{6.5}$$

$$\mathrm{BIC} = m \times \ln l - \ln L(x|M, \boldsymbol{\theta}), \tag{6.6}$$

where m is the number of the unknown parameters; l is the length of the observed data sample; $L(x|M, \boldsymbol{\theta})$ is the maximum value of the likelihood function of the fitted model. The Gaussian component number is the optimal number which produce lowest AIC or BIC.

6.2.3 Fatigue Reliability Estimation Methods

Based on the estimated probability distributions of S_{eq} and the statistic information listed in Table 6.2, fatigue reliability can be obtained from either of the following two methods

(1) Method I. An explicit formula of the fatigue reliability index β can be derived when the lognormal distribution is adopted for the daily S_{eq}. For a variable x that follows a lognormal distribution, the probability density function is

$$f(x) = \begin{cases} 0 & x \le 0 \\ \frac{1}{\sqrt{2\pi}\sigma x} \mathrm{e}^{-\frac{1}{2}\left(\frac{\ln x - \mu}{\sigma}\right)^2} & x > 0 \end{cases} \tag{6.7}$$

The mean value $\mu_{\ln X}$ and the standard deviation $\sigma_{\ln X}$ of the variable ln$X$ can be expressed as,

$$\begin{cases} \mu_{\ln X} = \ln\left(\frac{\mu_X}{\sqrt{1+\delta_X^2}}\right) \\ \sigma_{\ln X} = \sqrt{\ln\left(1+\delta_X^2\right)} \end{cases}, \tag{6.8}$$

where $\delta_X = \sigma_X/\mu_X$. Assuming that all the variables in Eq. (6.1) are independent, the fatigue reliability index β can be defined as

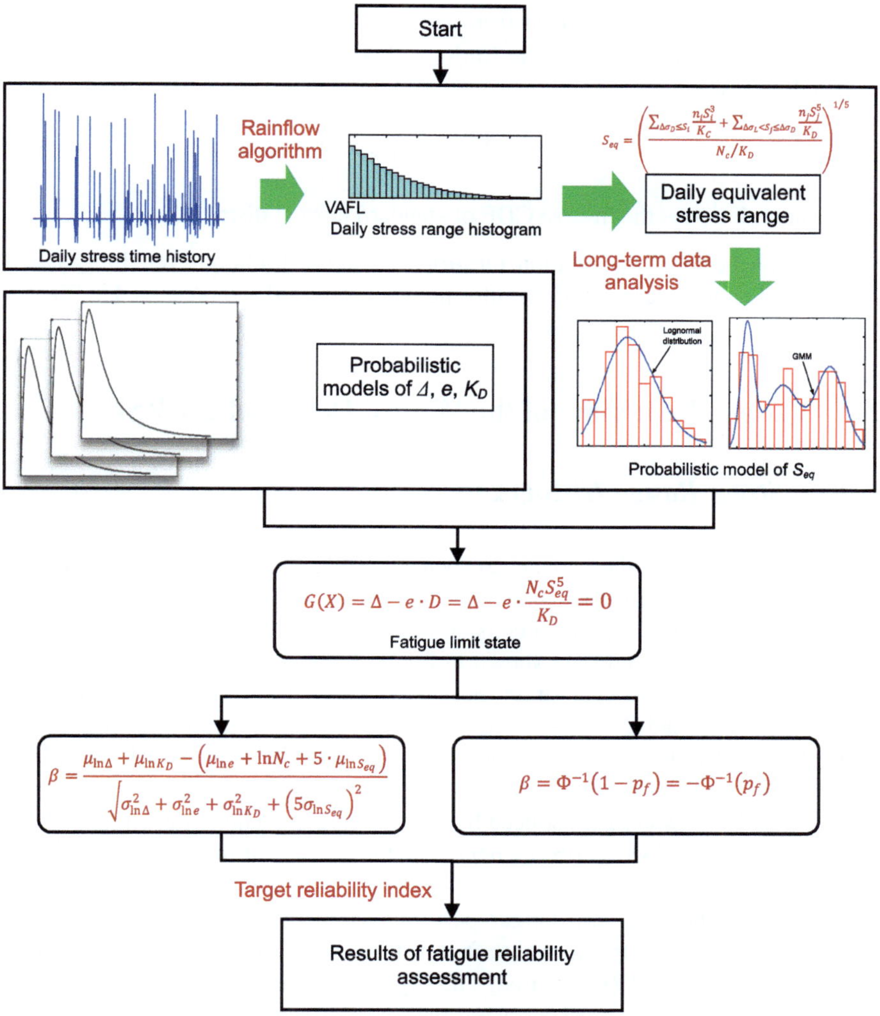

Fig. 6.1 Outline of fatigue reliability assessment based on long-term strain monitoring

$$\beta = \frac{\sum \mu_{\ln X_i,R} - \sum \mu_{\ln X_i,S}}{\sqrt{\sum \sigma_{\ln X_i,R}^2 + \sum \sigma_{\ln X_i,S}^2}} = \frac{\mu_{\ln \Delta} + \mu_{\ln K_D} - \left(\mu_{\ln e} + \ln N_c + 5 \cdot \mu_{\ln S_{\mathrm{eq}}}\right)}{\sqrt{\sigma_{\ln \Delta}^2 + \sigma_{\ln e}^2 + \sigma_{\ln K_D}^2 + \left(5\sigma_{\ln S_{\mathrm{eq}}}\right)^2}}, \tag{6.9}$$

where $\mu_{\ln X}$ and $\sigma_{\ln X}$ denote the mean value and the standard deviation of $\ln X$ (herein, $X = \Delta$, e, K_D or S_{eq}), respectively.

Method II. The fatigue failure probability can be calculated by using the Monte Carlo method due to the difficulty in developing an explicit formula for the fatigue reliability index with the fitted GMM of S_{eq}. Instead, the fatigue reliability index can be derived from the fatigue failure probability p_f, which is simulated by using the Monte Carlo method. Therefore, the fatigue reliability index β is

$$\beta = \Phi^{-1}(1 - p_f) = -\Phi^{-1}(p_f) \tag{6.10}$$

where $\Phi^{-1}(\cdot)$ is the inverse CDF of standard normal distribution.

Based on the above formulation, the proposed outline for fatigue reliability assessment based on long-term strain monitoring is summarized and presented in Fig. 6.1.

6.3 Fatigue Reliability of the Welded Details of the RSB

6.3.1 Stress Range Histograms

The prototype bridge is the Runyang Suspension Bridge (RSB). The strain monitoring system has been briefed in Chap. 5.

The dynamic strain measurements collected by the strain gauges are processed and analyzed in this study. As shown in Fig. 6.2, gauges ZLNL4-13 and ZLNL4-15 are used for longitudinal strain measurements of the rib-to-deck weld details, while ZLNL4-14 and ZLNL4-16 are measuring transverse strains of the rib-to-rib weld details. The thicknesses of the deck plate and the U-shape rib are 14 and 6 mm, respectively. The rib-to-deck fillet welds were shop welded to obtain 100% penetration. U ribs were field spliced by butt welding in conjunction with two back-up plates and embedded U rib segment. The thicknesses of the back-up plates and the embedded U rib are both 6 mm.

To evaluate the long-term fatigue performance of the rib-to-deck and rib-to-rib weld details subject to operational traffic loadings, monitoring data collected during a total of 327 days in 2009 are used. The stress range histograms from the 327-day monitoring data are obtained by using the rain-flow algorithm [25], as presented in Fig. 6.3. It can be observed that small-amplitude stress ranges less than 3 MPa dominate the histograms, which is a typical finding for fatigue details of highway steel bridge [2, 14, 19]. Table 6.3 lists the number of stress cycles in the histograms, which indicates that these details have similar number of total cycles. The numbers of stress cycles, which are larger than the cut-off limit or the VAFL, only accounts for a small portion of the total cycles. Furthermore, the cycle number of the welded details in the downstream is less than that of the welded details in the upstream. This phenomenon may be attributed to the differences in the traffic volumes between upstream and downstream directions.

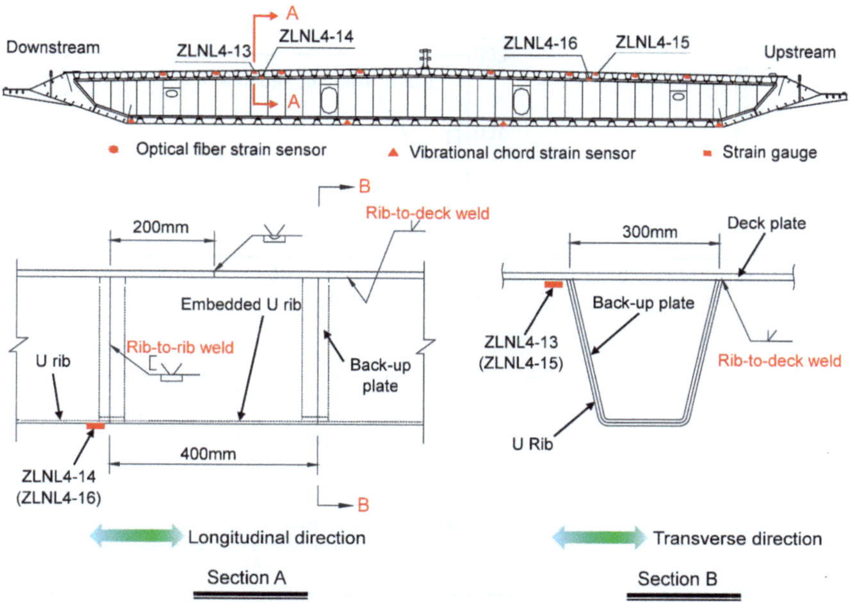

Fig. 6.2 Layout of strain sensors at the mid-span

Table 6.3 Cycles of stress ranges

Welded details	Total cycle number	Cycle number (larger than $\Delta\sigma_L$)	Cycle number (larger than $\Delta\sigma_D$)
ZLNL4-13	1.319×10^8	660.5	76.5
ZLNL4-14	1.384×10^8	143,039	2537.5
ZLNL4-15	1.276×10^8	17,620	826.5
ZLNL4-16	1.379×10^8	270,087	5468

6.3.2 *Probability Density Functions*

As introduced in Sect. 6.2.2, the probability density function of S_{eq} for the four selected weld details can be obtained using the lognormal distribution and the GMM. First, daily stress range histograms can be derived from recorded stress time histories by the rain-flow counting. Then the S_{eq} can be obtained using Eq. (5.8). A total of 327 data points of S_{eq} are obtained from 327-day measurements, which are further used to develop the probability density function of S_{eq}. Figures 6.4 and 6.6 plot the statistical histograms of S_{eq} for the four details. As can be seen, for the rib-to-deck details, there is only one dominant peak. Therefore, the lognormal distribution can be fitted to these data points using the maximum likelihood method. The fitted probability density functions of S_{eq} for the two rib-to-deck details are presented in Eqs. (6.11) and (6.12).

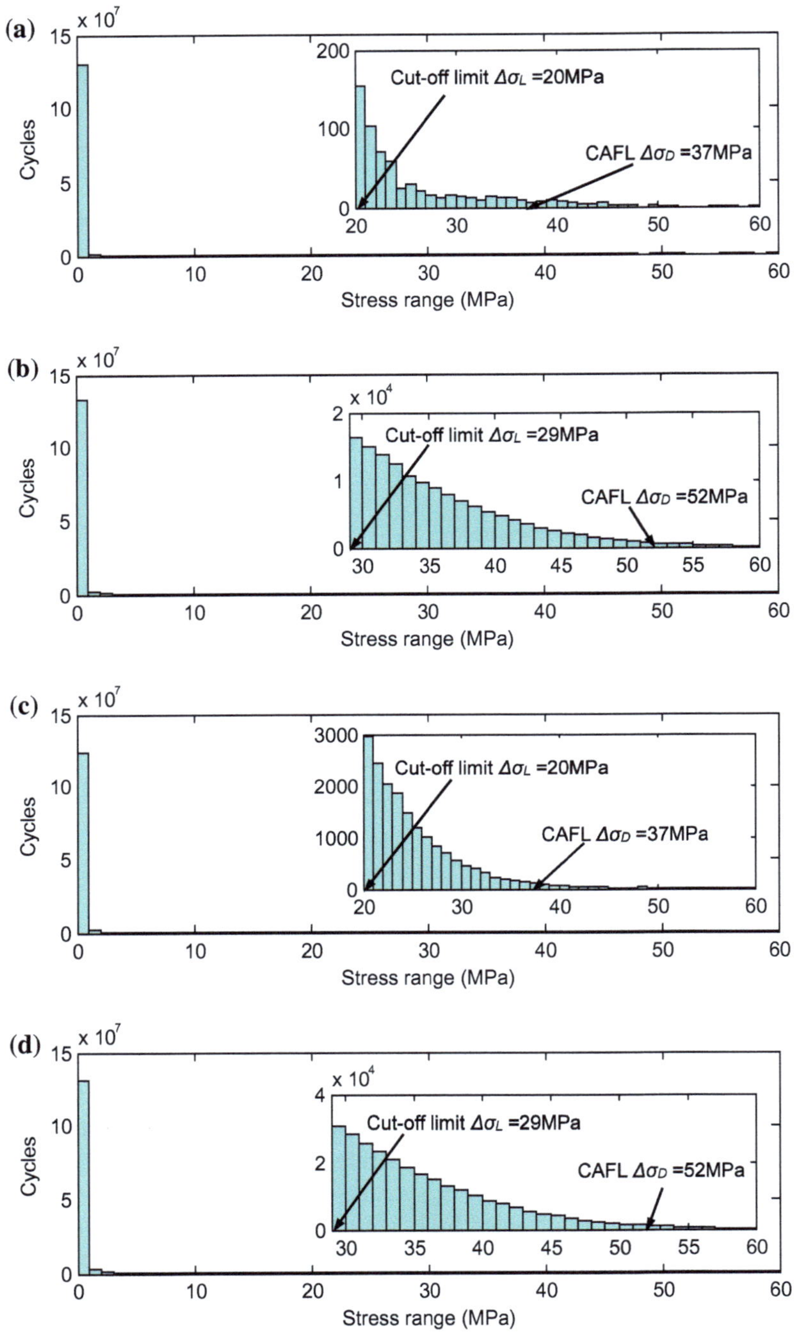

Fig. 6.3 Stress range histograms: **a** ZLNL 4-13, **b** ZLNL 4-14, **c** ZLNL 4-15 and **d** ZLNL 4-16

Fig. 6.4 Probability densities of Seq of the rib-to-deck details: **a** ZLNL 4-13 and **b** ZLNL 4-15

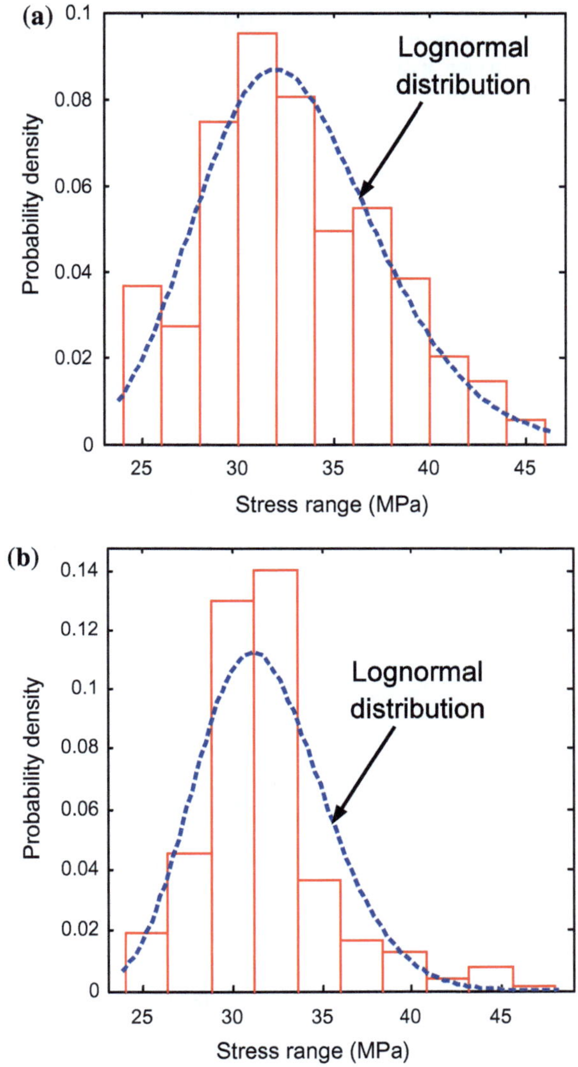

For the rib-to-rib details, there exist multi-peaks in the S_{eq} histograms, and thus the GMM is used. The optimal number of Gaussian components is firstly determined by using the information criterions introduced in Sect. 6.2.2. Figure 6.5 shows the values of AIC and BIC with the number of Gaussian components varying from 1 to 10. It is observed that both the AIC and BIC are minimized when the number of Gaussian components is 3, meaning that the optimal number of Gaussian components for both ZLNL4-14 and ZLNL4-16 is 3. The estimated parameters are shown in Table 6.4 (Fig. 6.6).

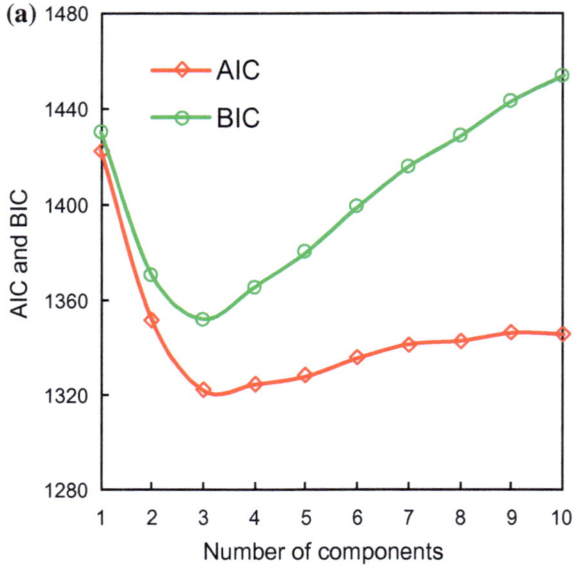

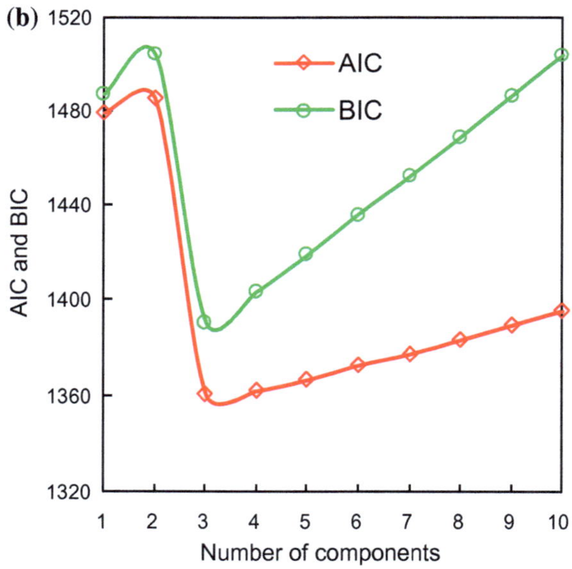

Fig. 6.5 Criterions for GMM component determination: **a** ZLNL 4-14; **b** ZLNL 4-16

$$\text{For ZLNL4}-13, f\left(S_{\text{eq}}\right) = \frac{1}{\sqrt{2\pi} \times 0.142 \times S_{\text{eq}}} e^{-\frac{1}{2}\left(\frac{\ln S_{\text{eq}}-3.48}{0.142}\right)^2} \quad S_{\text{eq}} > 0 \quad (6.11)$$

$$\text{For ZLNL4}-15, f\left(S_{\text{eq}}\right) = \frac{1}{\sqrt{2\pi} \times 0.113 \times S_{\text{eq}}} e^{-\frac{1}{2}\left(\frac{\ln S_{\text{eq}}-3.45}{0.113}\right)^2} \quad S_{\text{eq}} > 0 \quad (6.12)$$

Table 6.4 GMM parameters for the equivalent stress range S_{eq}

Component i	ZLNL 4-14			ZLNL 4-16		
	w_i	μ_i	σ_i^2	w_i	μ_i	σ_i^2
1	0.306	39.0	0.54	0.262	32.8	0.23
2	0.236	33.3	0.37	0.346	35.4	1.48
3	0.458	36.1	3.23	0.392	39.0	1.15

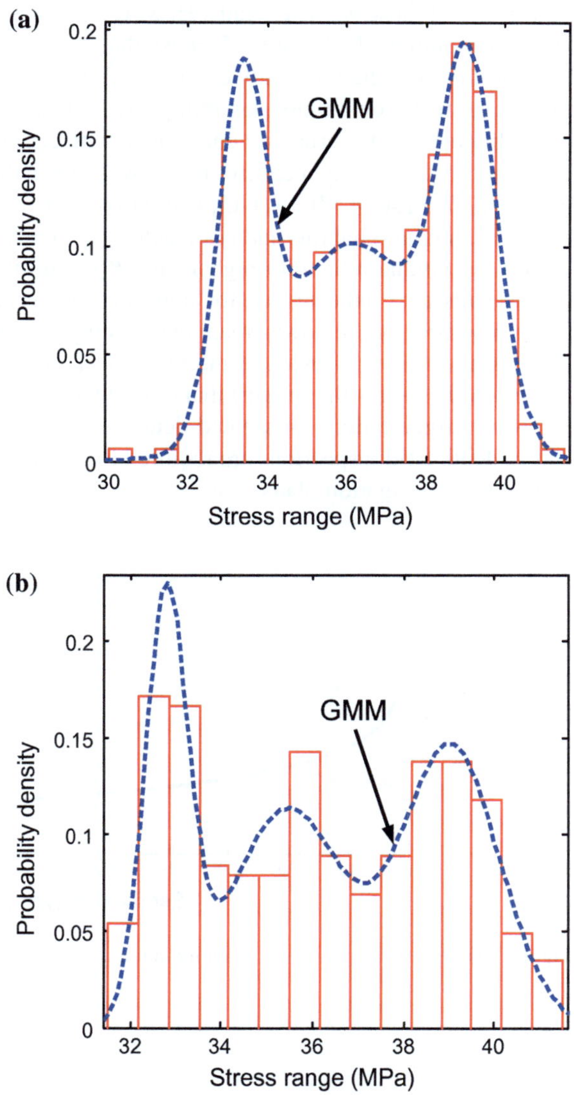

Fig. 6.6 Probability densities of S_{eq} of the rib-to-rib details: **a** ZLNL 4-14; **b** ZLNL 4-16

6.3.3 Reliability Results

With the statistic information listed in Table 6.2 and the probability density functions of S_{eq}, the fatigue reliabilities of the rib-to-deck details and the rib-to-rib details can be calculated using Eqs. (6.9) and (6.10) (using the Monte Carlo method). It is noted that the times of the Monte Carlo simulation is set to be 2×10^8.

The time-dependent fatigue reliability indices are shown in Figs. 6.7 and 6.8. It is observed that the reliability indices are pretty high at the beginning of the service life of these welded details. However, as time goes on, the reliability indices decrease significantly. Figure 6.7 shows that during the first 10 years of the service life, the fatigue reliability indices of the rib-to-rib details cannot be obtained due to the extremely low failure possibility. According to the ISO 2394 [26], the target reliability index for a structural component with certain consequences of failure and moderate costs of safety measures is 2.3. As show in the figures, the service years for details of ZLNL4-14, ZLNL4-15, and ZLNL4-16 are 33, 65, and 18, respectively, when the fatigue reliability indices reach β_{target}. Only the welded detail of ZLNL4-13 has the fatigue reliability higher than β_{target} during the entire 100-year service life. Results also indicate that for both the two types of weld details, the fatigue reliability index of the downstream detail is higher than that of the upstream detail. Besides, the reliability index of the rib-to-deck detail is higher than that of the rib-to-rib detail. The results suggest that during the service life the RSB, there are high levels of fatigue failure for welded details of the orthotropic deck. Hence, routine and in-depth field inspections are required to locate the potential fatigue cracks and make timely mitigation plans.

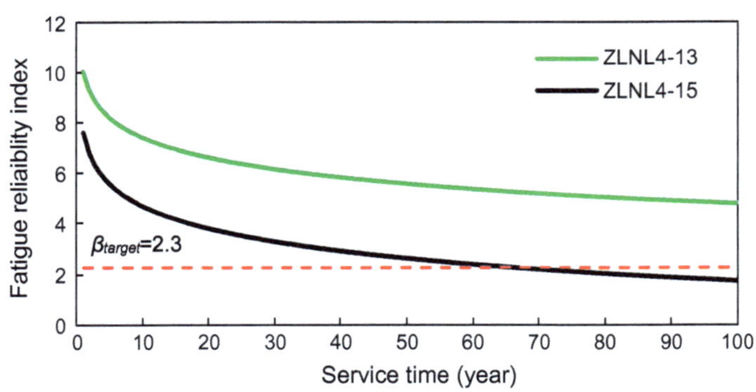

Fig. 6.7 Time-dependent fatigue reliability indices of the rib-to-deck welds

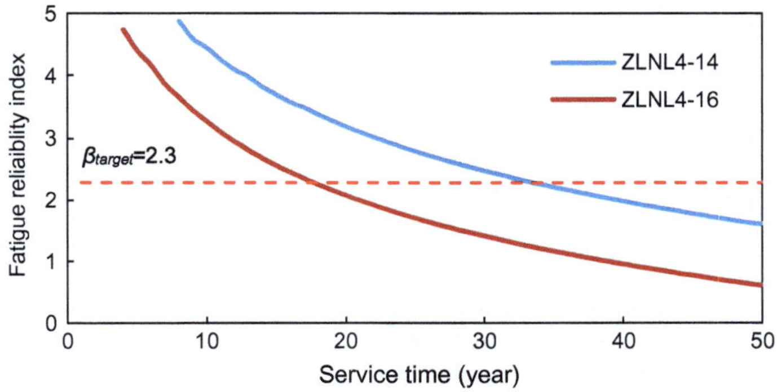

Fig. 6.8 Time-dependent fatigue reliability indices of the rib-to-rib welds

6.4 Summary

This study proposed a time-dependent fatigue reliability assessment approach for welded details of OSDs using long-term strain monitoring data. According to the methodologies, fatigue reliabilities of two types of welded details (i.e., rib-to-deck and rib-to-rib details) of an in-service long-span suspension bridge are investigated. Following conclusions can be made:

1. Two probabilistic models, namely, the lognormal distribution and the GMM, are adopted to quantify uncertainties of the daily S_{eq}. The lognormal distribution is more suitable for the unimodal Seq for rib-to-deck details. By contrast, the daily Seq for the rib-to-rib details can be represented by the GMM, which is composed of 3 Gaussian components.
2. The fatigue reliability can be calculated based on either an explicit formula or the Monte Carlo method. The results indicate that the reliability indices decrease significantly as the service life increases. During the 100-year service life, except for a rib-to-deck detail, other three welded details cannot meet the target fatigue reliability during the bridge's 100-year service life.
3. This study also reveals that the fatigue reliability indices of the downstream details are higher than those of the upstream details, which is probably due to the difference in the traffic volumes between upstream and downstream directions. Besides, the rib-to-deck details for the RSB have higher fatigue reliabilities than those of the rib-to-rib details.

The proposed approach can help bridge owners and practitioners make informed decisions regarding maintenance and repair of potential fatigue cracks.

References

1. Wolchuk R. Lessons from weld cracks in orthotropic decks on three European bridges. J Struct Eng. 1990;116(1):75–84.
2. Deng Y, Liu Y, Feng DM, Li AQ. Investigation of fatigue performance of welded details in long-span steel bridges using long-term monitoring strain data. Struct Control Health Monit. 2015;22(11):1343–58.
3. Xiao ZG, Yamada K, Inoue J, Yamaguchi K. Fatigue cracks in longitudinal ribs of steel orthotropic deck. Int J Fatigue. 2006;28(4):409–16.
4. Fisher JW, Barsom JM. Evaluation of cracking in the rib-to-deck welds of the Bronx-Whitestone bridge. J Bridge Eng. 2016;21(3):04015065.
5. British Standards Institution. Eurocode 3: design of steel structures—Part 1–9. Fatigue;2005.
6. Zhang QH, Cui C, Bu YZ, Liu YM, Ye HW. Fatigue tests and fatigue assessment approaches for rib-to-diaphragm in steel orthotropic decks. J Constr Steel Res. 2015;114:110–8.
7. Kainuma S, Yanga M, Jeong YS, Inokuchi S, Kawabata A, Uchida D. Experiment on fatigue behavior of rib-to-deck weld root in orthotropic steel decks. J Constr Steel Res 2016;119:113–22.
8. Guo T, Li AQ, Li JH. Fatigue life prediction of welded joints in orthotropic steel decks considering temperature effect and increasing traffic flow. Struct Health Monit Int J. 2008;7(3):189–202.
9. De Freitas ST, Kolstein H, Bijlaard F. Structural monitoring of a strengthened orthotropic steel bridge deck using strain data. Struct Health Monit Int J. 2012;11(5):558–76.
10. Guo T, Liu ZX, Zhu JS. Fatigue reliability assessment of orthotropic steel bridge decks based on probabilistic multi-scale finite element analysis. Adv Steel Const. 2015;11(3):334–46.
11. Liu Y, Zhang H, Liu Y, Deng Y, Jiang N, Lu N. Fatigue reliability assessment for orthotropic steel deck details under traffic flow and temperature loading. Eng Fail Anal. 2017;71:179–94.
12. Frangopol DM, Strauss A, Kim S. Bridge reliability assessment based on monitoring. J Bridge Eng. 2008;13(3):258–70.
13. Liu M, Frangopol DM, Kwon K. Fatigue reliability assessment of retrofitted steel bridges integrating monitored data. Struct Saf. 2010;32(1):77–89.
14. Wirsching PH. Fatigue reliability for offshore structures. J Struct Eng ASCE. 1984;110(10):2340–56.
15. Zhao ZW, Haldar A, Breen FL. Fatigue-reliability evaluation of steel bridges. J Struct Eng ASCE. 1994;120(5):1608–23.
16. Kwon K, Frangopol DM. Bridge fatigue reliability assessment using probability density functions of equivalent stress range based on field monitoring data. Int J Fatigue. 2010;32(8):1221–32.
17. Guo T, Chen YW. Field stress/displacement monitoring and fatigue reliability assessment of retrofitted steel bridge details. Eng Fail Anal. 2011;18(1):354–63.
18. Deng Y, Ding YL, Li AQ, Zhou GD. Fatigue reliability assessment for bridge welded details using long-term monitoring data. Sci China Technol Sci. 2011;54(12):3371–81.
19. Ni YQ, Ye XW, Ko JM. Monitoring-based fatigue reliability assessment of steel bridges: analytical model and application. J Struct Eng. 2010;136(12):1563–73.
20. Xia HW, Ni YQ, Wong KY, Ko JM. Reliability-based condition assessment of in-service bridges using mixture distribution models. Comput Struct. 2012;106:204–13.
21. Titterington DM, Smith AFM, Makov HE. Statistical analysis of finite mixture distributions. Chichester: Wiley; 1985.

22. Dempster AP, Laird NM, Rubin DB. Maximum likelihood from incomplete data via the EM algorithm. J R Stat Soc Ser B (Methodological). 1977;39(1):1–38.
23. Akaike H. A new look at the statistical model identification. IEEE Trans Autom Control. 1974;19(6):716–23.
24. Schwarz G. Estimating the dimension of a model. Ann Stat. 1978;6(2):461–4.
25. Nieslony A. Determination of fragments of multiaxial service loading strongly influencing the fatigue of machine components. Mech Syst Signal Process. 2009;23(8):2712–21.
26. International Organization for Standardization. ISO 2394: general principles on reliability for structures: 1996.

Part 4

Chapter 7
Vertical Deflection Monitoring of Main Girder

7.1 Introduction

Long-span bridges play an important role in the people's daily life and transportation. It is necessary to establish a reliable health monitoring system to maintain the function and safety of long-span bridges. In long-span bridges, monitoring of the vertical deflection under the actual traffic loads is not only crucial to ensure the overall serviceability and safety of the bridge, but also to predict the abnormal state. Micrometers or displacement meters are widely used to measure bridge deflections directly in evaluation of old field bridges and tests of newly built bridge [1]. For long-term deflection monitoring, more automatic and convenient displacement sensors are needed [2].

Global positioning system (GPS) has been becoming a popular technique for the deflection monitoring of long-span bridges. Yi et al. [3, 4] conducted extensive research on the characterization and extraction of the GPS multipath signals and the performance assessment of high-rate GPS receivers. Ogundipe et al. [5] used the GPS to measure three main frequencies of a steel box girder viaduct. The results showed that deflections can be measured at the millimeter level. Although using GPS to measure bridges' displacements has several advantages, some problems still remain to be solved. These include the high level of electromagnetic noise, usage limitation due to the weather and high cost [6].

Some non-directional sensor systems have been invented for deflection monitoring, such as inclinometers and long-gauge fiber optic sensors [1, 7, 8]. Chung et al. [7] placed two pairs of 3 m long-gauge sensors on a prestressed concrete girder with a parallel topology. He et al. [8] designed an inclinometer and then developed a reliable integration method to measure the deflection of a high-speed railway bridge. In these systems the bridge deflections are often estimated from the other structural responses using the integration algorithm or classical beam theory. However, it is not easy to find a proper relationship between indirect measurements such as curvature and bridge deflection for complex bridge systems, such as cable-stayed bridges and

Y. Deng and A. Li, *Structural Health Monitoring for Suspension Bridges*,
https://doi.org/10.1007/978-981-13-3347-7_7

suspension bridges. With the development of photonics science and image processing technology, vision-based monitoring systems have been developed as an effective alternative for the displacement or deflection monitoring of bridges [2, 9, 10]. Ye et al. [2] developed a vision-based deflection system consisting of a high-resolution industrial CCD digital camera and an extended-range zoom lens. Kohut et al. [9] developed a vision-based deflection system including one or more high-resolution digital cameras. The previous studies indicated that the vision-based monitoring systems needed a complex algorithm to derive the actual displacement and can be influenced by rainy or foggy weather.

Although considerable work has been carried out on structural displacement or deflection monitoring, developing a high-reliability, cost-effective, and real-time system with the characteristics of long-distance and high-precision for the deflection monitoring of long-span bridges is still desirable. For this reason, some work on the CPS has been carried out in recent years [11–14]. The basic concept of the CPS is that the reference point (fixed) can be connected with the measurement points using pipes filled with liquid, then the changes of the relative positions between the reference point and the measurement points will lead to the changes of the liquid levels in the connected pipes. Hence, the changes of the liquid levels detected by some types of sensors can be used to deduce the position changes of the measurement points. The CPS had been used in the health monitoring of the Dongting Lake Bridge (a three-tower cable-stayed bridge) [11], Dafoshi Yangtze River Bridge (a two-tower cable-stayed bridge) [12, 13] and Jing-yue Yangtze River Highway Bridge (a two-tower cable-stayed bridge) [14] in China.

So far, most studies focused on developing novel and accurate monitoring techniques for deformation or deflection. Actually, effective application of monitored data on performance assessment for engineering structures is as important as inventions of the new monitoring techniques. Some studies began to utilize the deflection monitoring data to identify the dynamic characteristics of engineering structures [5, 15–19]. GPS-based deflection monitoring data were often used to identify the modal frequencies and shapes [5, 15, 16]. Besides, some new deflection monitoring techniques such as robotic theodolite [17], interferometric radar [18, 19] were invented and used to monitor the dynamic displacement and then identify the dynamic characteristics of engineering structures. At present, it is still a challenge that how to use the monitored deflection data to assess bridges' performance fully.

This chapter presents a CPS that is particularly designed and established for the deflection measurement of the NSB, using pressure transmitters to detect the changes of liquid levels in the connected pipes. The deflections can be deduced using a simple formula from the changes of liquid levels. The performance of the developed CPS is verified through both controlled and uncontrolled load tests. In addition, this chapter attempts to propose an alternative of deflection assessment method using long-term monitored data for long-span bridges. A deflection threshold system including two levels is determined. With the long-term monitored data, the extreme deflections in the design reference period is predicted by the means of EVA and then compared with the thresholds.

7.2 Vertical Deflection Monitoring System of the NSB

7.2.1 Connected Pipe System

Designing a proper deflection monitoring system for long-span bridges is an important aspect in bridge structural health monitoring. Pressure transmitters were adopted as the sensor to detect the changes of liquid levels as shown in Fig. 7.1. When this system is filled with water, the valve is open. After the air in the connected pipes and the sensors is completely vented out, this valve is closed. In this figure, the reference point is located at the river bank or at the pier. The measurement points can be line-likely distributed to monitor the bridge deflection and then track the profile of the bridge under the external loads. The principle of deflection monitoring using the CPS with pressure transmitters can be seen with Fig. 7.1.

At the beginning (time = 0), the pressure of the reference point $P_{1,0}$ is detected by the pressure transmitter as

$$P_{1,0} = \rho g H_{1,0} \tag{7.1}$$

where ρ = density of water; g = acceleration of gravity at the site of bridge and $H_{1,0}$ = level difference from the reference point to the water surface. The pressures of the ith measurement point $P_{i,0}$ is:

$$P_{i,0} = \rho g H_{i,0}, \tag{7.2}$$

where $H_{i,\,0}$ = level difference from the ith measurement point to the water surface. At this moment, the height difference between the reference point and the ith measurement point $\Delta H_{i,\,0}$ is

Fig. 7.1 Schematic diagram of the CPS

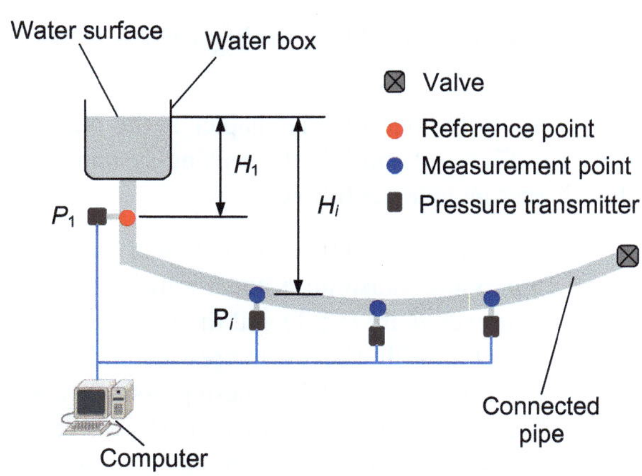

$$\Delta H_{i,0} = H_{i,0} - H_{1,0} = \frac{P_{i,0} - P_{1,0}}{\rho g} \tag{7.3}$$

After the external loads have been imposed to the bridge (time = T), the bridge deformed and the pressures of both reference point and measurement points change to

$$P_{1,T} = \rho g H_{1,T} \tag{7.4a}$$

$$P_{i,T} = \rho g H_{i,T} \tag{7.4b}$$

The height difference between the reference point and the ith measurement point has also changed

$$\Delta H_{i,T} = H_{i,T} - H_{1,T} = \frac{P_{i,T} - P_{1,T}}{\rho g} \tag{7.5}$$

Hence, at time = T the current deflection of the ith measurement point relative to the beginning (time = 0) is

$$D_{i,T} = \Delta H_{i,T} - \Delta H_{i,0} = \frac{P_{i,T} - P_{1,T} - P_{i,0} + P_{1,0}}{\rho g} \tag{7.6}$$

The concept and principle of the CPS for deflection monitoring described above is simple and easy to realize. In addition, it should be noted that choosing proper pressure transmitters with enough range can allow the CPS suitable to monitor the long-span bridges with large longitudinal gradient, such as cable-stayed bridges and suspension bridges. This system can easily track the profile of the bridge's main girder with distributed measurement points.

7.2.2 Deflection Monitoring of the NSB

The prototype bridge of this chapter is the Nanxi Suspension Bridge (NSB). As shown in Figs. 7.2 and 7.3, the sensing systems in SHMS of the NSB involved in this chapter include the following:

1. CPS to monitor the vertical deflection of the steel box girder. This system consists of 15 pressure transmitters and a water box. A pressure transmitter representing the reference point (SYT) and the water box were installed in the Yibin tower, shown in Fig. 7.2b. The rest 14 pressure transmitters were installed in the steel box girder, shown in Fig. 7.2a. These pressure transmitters were located at sections of $l/8$, $2l/8$, $3l/8$, $l/2$, $5l/8$, $6l/8$, and $7l/8$, where l is the main span of the NSB. The sample frequency of pressure transmitters was set to be 0.5 Hz. The accuracy of the pressure transmitter is 0.04% of the measuring span. The maximum span of

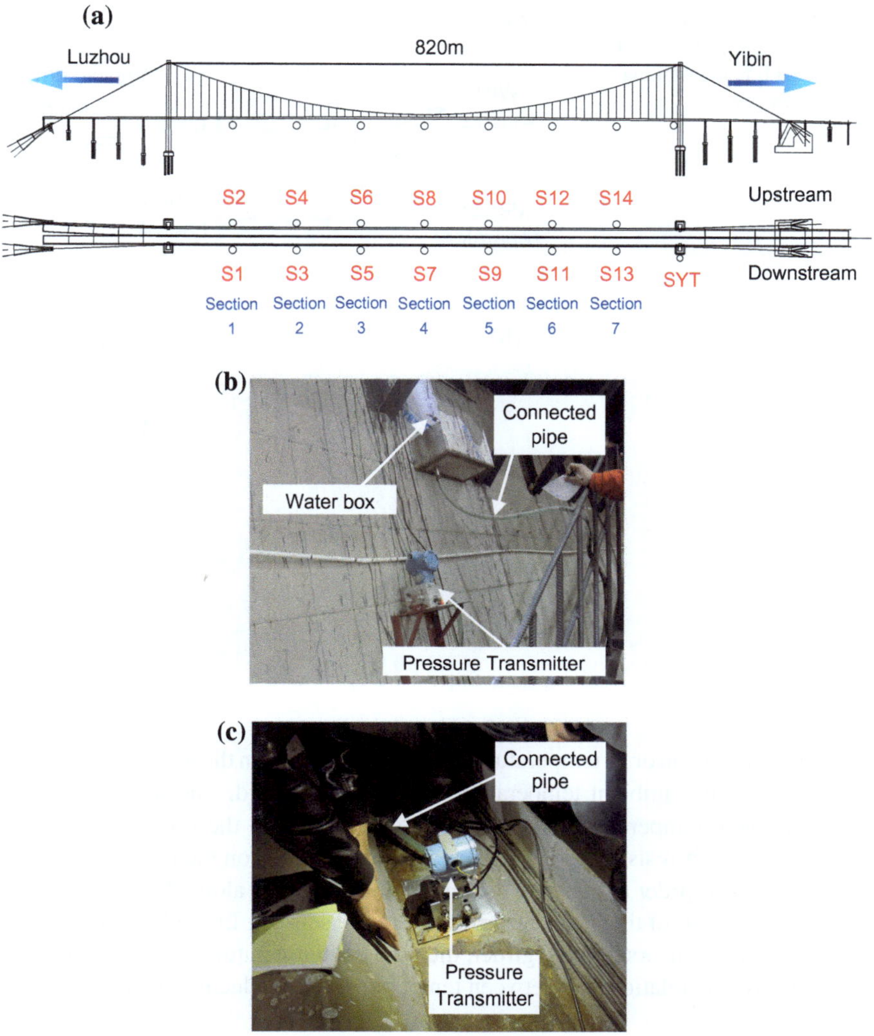

Fig. 7.2 CPS of the NSB: **a** arrangement of pressure transmitters, **b** CPS facilities in the Yibin tower and **c** CPS facilities in the steel box girder

the transmitter is 2000 psi (13.8 MPa). The span ratio (the maximum span to the minimum span) is 150:1.

2. Weigh-in-motion (WIM) system. The WIM sensors were installed in the pavement at the location of the Luzhou tower, shown in Fig. 7.3. There were four WIM sensors in lane 1, lane 2, lane 3, and lane 4 to monitor the vehicle load, respectively. While lane 1 and lane 2 carry the vehicles driving from the Yibin city to the Luzhou city, lane 3 and lane 4 carry the vehicles driving in the opposite direction.

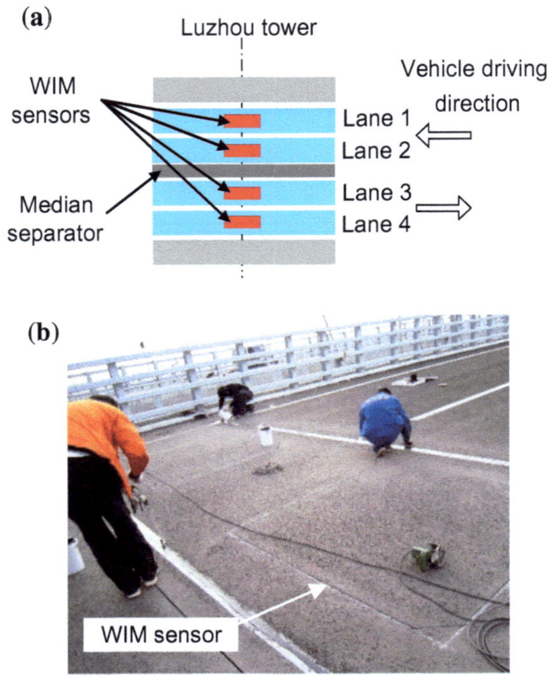

Fig. 7.3 WIM system of the NSB: **a** arrangement of WIM sensors and **b** field installing of WIM sensors

3. Temperature sensor. A temperature sensor was installed in the middle of the girder to monitor the ambient temperature, shown in Fig. 1.1d. The sensor measures the ambient temperature once per minute. Actually, in the middle of the span there are 15 Pt-resistance temperature sensors installed on the medial surface of the steel box girder to measure the temperature gradient along the depth. Figure 1.1c shows part of the Pt-resistance temperature sensors. Instead of Pt-resistance temperature sensors in main girder, the ambient temperature sensor will be used to discuss the relationship between temperature and deflection qualitatively.

7.2.3 Data Analysis Framework

With the sensors mentioned above, the deflection assessment for the NSB mainly includes following steps:

Step 1: Extracting the vehicle-induced deflection from the data monitored by CPS. The vertical deflection of a suspension bridge is sensitive to the changes of vehicle loads and ambient temperature. In a particular area, the changes of ambient temperature are relatively deterministic when they are compared with the changes of vehicle loads. However, there is a significant randomness in the actual vehicle loads. And sometimes the traffic weight and flow may increase. Thus, the temperature-induced deflection should be eliminated from the original deflection data. Then the extracted vehicle-induced deflection will be used to check the serviceability and safety of the bridge.

Step 2: Verification of CPS. As a new technique of deflection monitoring, the verification is important to ensure the accuracy and reliability of CPS in long-term deflection monitoring. In the controlled load test, 14 three-axle trucks were placed on the deck of the main girder. The deflections recoded by CPS will be compared with the results finite element analysis. In the uncontrolled load test, the passage of a six-axle on the deck was recorded by WIM and CPS. This passage will be simulated by using the finite element transient analysis.

Step 3: Determination of deflection thresholds related to vehicle loads. In this step, the deflection threshold is determined without consideration of temperature for a couple of reasons. First, the design specification for highway suspension bridges (DSHSB) in China [20] gives the allowable deflection related to vehicle loads only. In addition, the General Code for Design of Highway Bridges and Culverts (GCDHBC) in China [21] only specifies the uniform temperature action rather than the details of the temperature gradient for the steel box girder. Hence, for the health monitoring of a suspension bridge under operation, it is more suitable to construct a threshold system to assess the vehicle-induced deflection monitored by the SHMS.

Step 4: Extreme deflection prediction and assessment. In this step, the daily maximum and minimum vehicle-induced deflections in 100 days in the first half of 2014 will be identified at first. Then the GEVD in EVA is used to predict the extreme vehicle-induced deflections in design reference period of 100 years. At last, the predicted extreme deflections of the NSB are compared with the deflection thresholds determined in step 3.

7.3 Verification of the Connected Pipe System

7.3.1 Processing of Deflection Measurements

To validate the effectiveness of the CPS in deflection monitoring, both controlled and uncontrolled load tests were performed. As we know, the changes of environmental conditions including temperature and vehicle loads can lead to the bridge's deformation. Furthermore, measuring noise can also affect the deflection monitoring results. Hence, it is important to investigate whether this CPS can trace the real change of deflection induced by the changes of environmental conditions. For this reason, the data measured on February 14, 2014 is selected and processed. Figure 7.4 gives the original deflection history measured by pressure transmitter S8 (in the upstream middle of the span). Figure 7.5 gives the original ambient temperature history. Herein presents a qualitative discussion of the relationship between temperature and deflection.

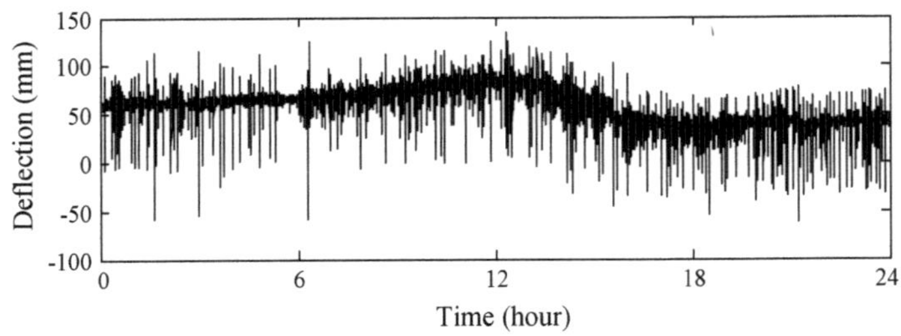

Fig. 7.4 Vertical deflection of section *l*/2 on February 14, 2014

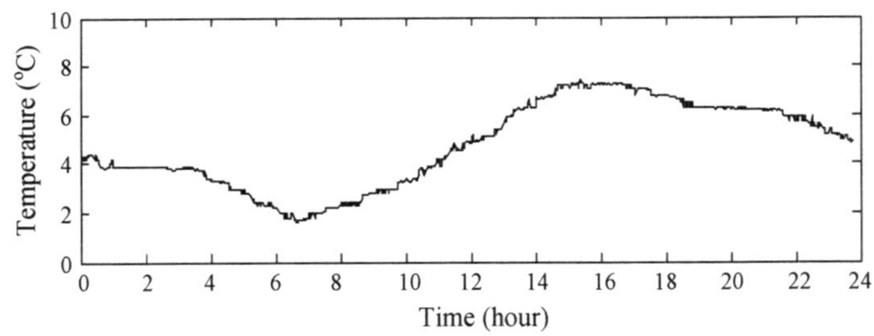

Fig. 7.5 Environmental temperature on February 14, 2014

In fact, three components of the vertical deflection can be derived from the original data in Fig. 7.4 as discussed below:

1. Temperature-induced deflection. It can be observed that the change pattern of temperature is large-small-large-small in Fig. 7.5. Conversely, on the whole the change pattern of the original vertical deflection is approximately small-large-small-large. Hence, there seems to be an inverse relationship between the vertical deflection and ambient temperature. This relationship was previously observed in a concrete box girder bridge [22] and a cable-stayed bridge [23]. Then, the wavelet transform is used to extract the deflection induced by the temperature fluctuation [24]. The results are shown in Fig. 7.6. In Fig. 7.5, the minimum and maximum temperatures appeared at about 7 o'clock and 15 o'clock, respectively. However, in Fig. 7.6 the maximum and minimum vertical deflections appeared at about 12 o'clock and 17 o'clock, respectively. Hence, the changing of the vertical deflection lags behind the temperature fluctuation.
2. Vehicle-induced deflection. Figure 7.7 shows the vertical deflection after removing the temperature effects. The deflection in this figure is considered to be induced mainly by vehicle loads and testing noise by ignoring the wind effects on the vertical deflection. It can be seen that the data points fluctuate around zero, indicating the temperature effects on the vertical deflection has been eliminated from the original deflection data efficiently. To examine the data points more clearly, a small data set measured from 0 o'clock to 9 min past 0 o'clock, derived from the vertical deflection after removing the temperature effects, is shown in

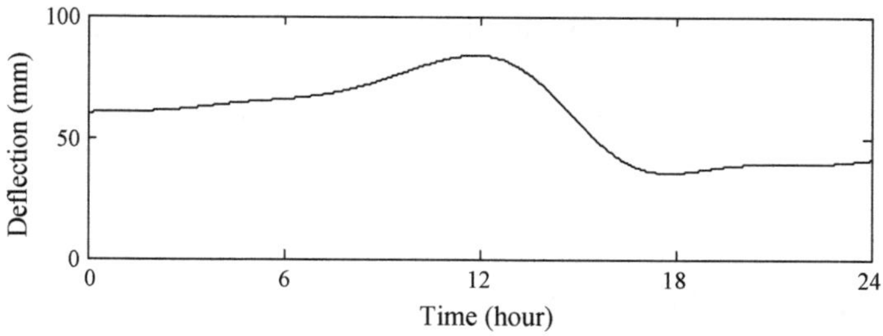

Fig. 7.6 Deflection induced by temperature of section *l*/2 on February 14, 2014

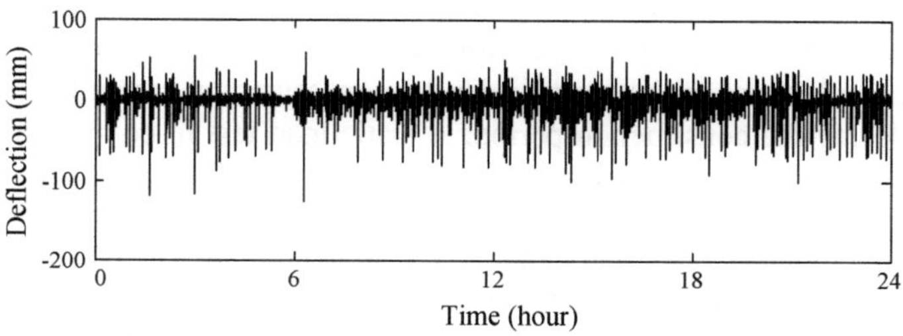

Fig. 7.7 Deflection after removing temperature effects of section *l*/2 on February 14, 2014

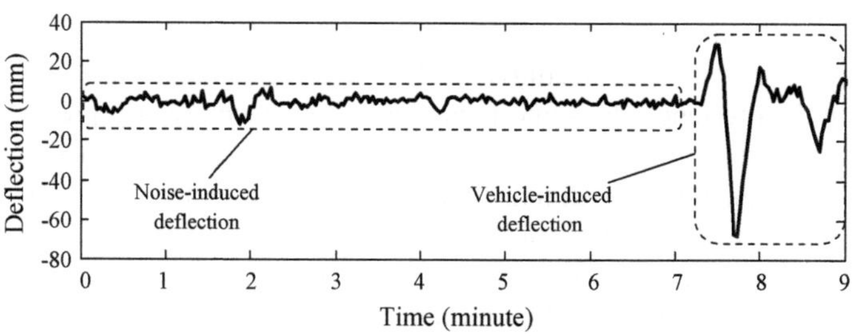

Fig. 7.8 Deflection after removing temperature effects from 00:00:00 to 00:09:00

Fig. 7.8. In Fig. 7.8 the data measured from 7 min past 0 o'clock to 9 min past 0 o'clock is caused by a 6-axle heavy truck. The maximum positive and minimum negative data points during this period are 28.9 mm and −68.3 mm, respectively. This deflection induced by a six-axle heavy truck will be further studied in the following uncontrolled load test.

3. Noise-induced deflection. Also shown in Fig. 7.8, from 0 o'clock to 7 min past 0 o'clock the data points have a random fluctuation. It is noted that during this period there may be some light vehicles on the bridge, which causes small deflec-

tions. It is almost impossible to separate this light vehicle-induced deflection from the measurement noise effects. Hence, the data points measured during this period are considered to be mainly induced by the measurement noise effects. The range of the noise-induced deflections is approximately from −10 to 10 mm.

7.3.2 Controlled Load Test

Before the NSB was open to the actual traffic, controlled load tests including static and dynamic tests were carried out to check the design work and make sure that the bridge could meet the needs in operation period. During the controlled load tests, the vertical deflections of the main girder were measured and recorded by the CPS installed during the construction period. A static load test in controlled load tests and finite element analysis were employed to verify the effectiveness of the CPS. It should be noted that the duration of the static load test was short so the temperature effects on the monitored vertical deflection were negligible. The original deflection data, measured after the test trucks were placed at the desired locations, will be compared with the finite element analysis results later. Figure 7.9 shows the testing truck placement of the selected static load test. There were 14 three-axle trucks and the gross weight of each truck was 300 kN. The axle weight distribution of each truck was 60, 120, and 120 kN.

In theory, when subjected to equivalent loads, the model of an accurately designed bridge should experience the same behavior as the structure it represents. On this principle, a finite element model, as seen in Fig. 7.10, was created in ANSYS software based on the design plans and dimensional/sectional parameters. The model was verified using results from the dynamic test. The measured first vertical modal frequency was 0.18 Hz. The calculated first vertical modal frequency was 0.15 Hz. The finite element model was not updated to eliminate this small error due to the uncertainties in the error's source. This controlled static load test was replicated in the model. The

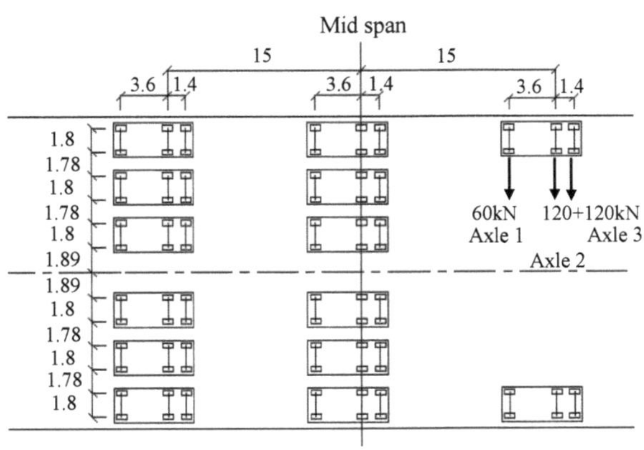

Fig. 7.9 Test truck deployment of a controlled load test (dimension unit in m)

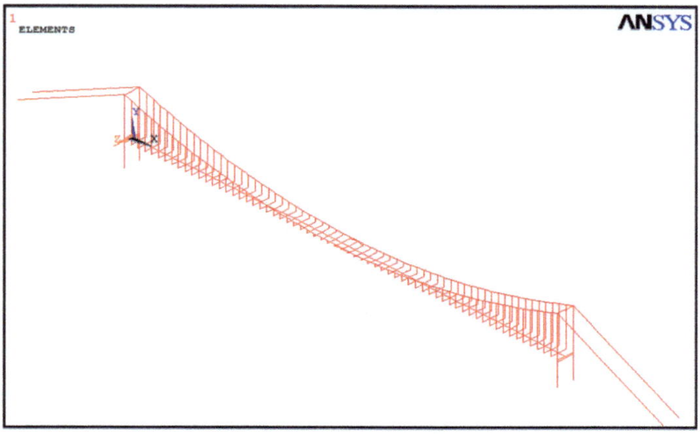

Fig. 7.10 Three-dimensional finite element model of NSB

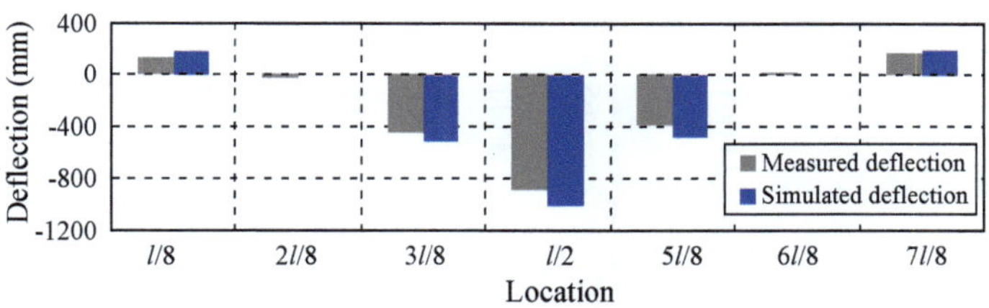

Fig. 7.11 Comparisons between the simulated results and the controlled load test data

measured and simulated vertical deflections were shown in Fig. 7.11. In the middle section, the measured and simulated deflections were −882 and −934 mm, respectively. It should be noted that the upward deflection is positive, and the downward deflection is negative. This difference in deflection of 52 mm, at a relative error of 5.9%, was considered acceptable. The comparisons prove that the CPS can measure the deflection of the main girder of the NSB fairly accurately in the static controlled load test.

7.3.3 Uncontrolled Load Test

After the controlled load tests, the NSB was open to the actual traffic. At the same time the CPS began to trace the girder's deflection in service condition. During this period the traffic loads are uncontrolled and random. Hence, it is necessary to check the capacity of the CPS to detect the changes of girder's deflection under the actual traffic. The data monitored in about 7 min is employed to validate the deflection monitoring capacity of the CPS under actual traffic loads.

Table 7.1 lists detailed information of a series of vehicles monitored by WIM system from 0:01:46 to 0:08:35 on February 14, 2014. There were nine vehicles

Table 7.1 A set of WIM data on February 14, 2014

Vehicle	Time of arrival	GVW (kN)	Speed (km/h)	Number of axle	Driving lane	Axle weight (kN)					
						1	2	3	4	5	6
Truck 1	0:01:46	97	67	2	Lane 1	32	65	/	/	/	/
Car 1	0:02:21	17	79	2	Lane 1	10	7	/	/	/	/
Car 2	0:02:49	24	73	2	Lane 2	13	11	/	/	/	/
Car 3	0:03:38	16	89	2	Lane 3	9	7	/	/	/	/
Truck 2	0:04:01	54	79	2	Lane 1	22	32	/	/	/	/
Truck 3	0:06:46	503	61	6	Lane 4	59	85	88	93	93	85
Car 4	0:07:03	18	91	2	Lane 3	9	9	/	/	/	/
Car 5	0:07:22	18	105	2	Lane 3	9	9	/	/	/	/
Truck 4	0:08:35	168	56	2	Lane 1	59	109	/	/	/	/

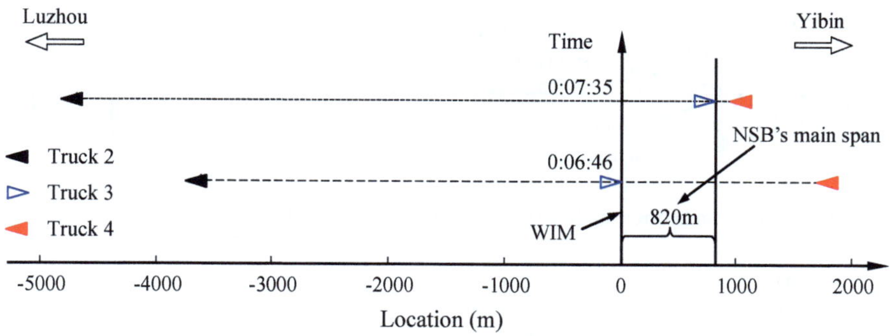

Fig. 7.12 Locations of trucks

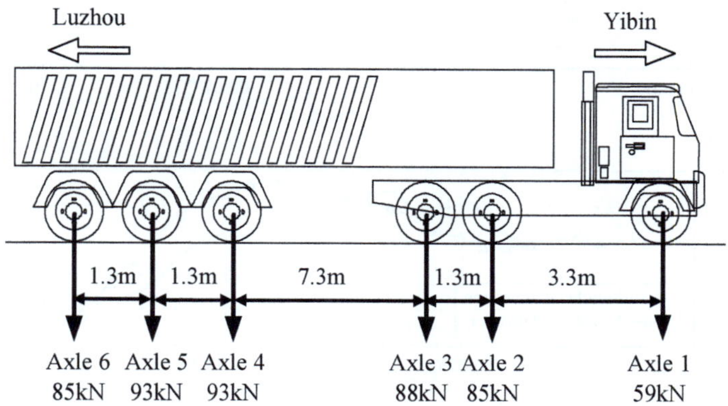

Fig. 7.13 Axle weight and spacing of Truck 3

including one six-axle truck, three two-axle trucks, and five two-axle cars. The gross vehicle weights (GVM) of all the cars were below 25 kN so the vertical deflection caused by these cars was apparently small and negligible. Hence, only the deflection caused by the trucks is considered. As listed in Table 7.1, there are times of arrival in the WIM sensors, driving speeds and driving lanes. Accordingly, the locations of Truck 2, Truck 3, and Truck 4 are determined and shown in Fig. 7.12. At 0:06:46 Truck 3 was captured by the WIM sensors. At the same time Truck 2 and Truck 4 were far from the NSB's main span. About 49 s later, at 0:07:35 Truck 3 was going to leave the NSB's main span. At this moment the locations of Truck 2 and Truck 4 were still out of the NSB's main span. Therefore, from 0:06:46 to 0:07:35 there was only Truck 3 on the deck of the NSB. Obviously during this period, the vertical deflection of the girder can be considered to be caused by Truck 3 only.

The crossing process of Truck 3 on the deck of the NSB is simulated by using the finite element transient analysis. The main girder is modeled by using the Beam4 elements, as illustrated in Fig. 7.10. The beam4 element has two nodes. Thus, Truck 3 is simplified as 6 concentrated forces, which are representing the six axles shown

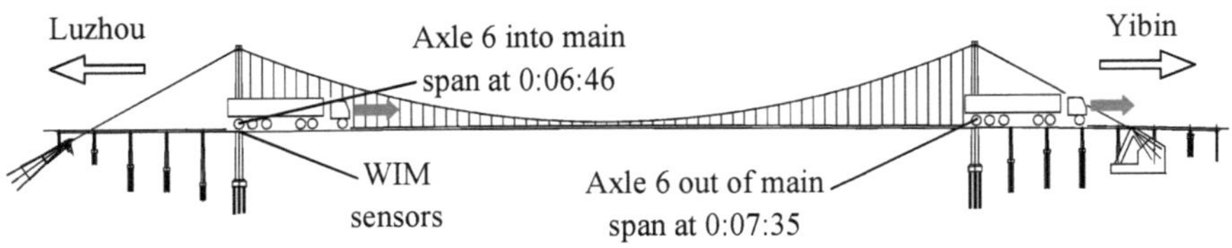

Fig. 7.14 Position and corresponding time of Truck 3 on the NSB

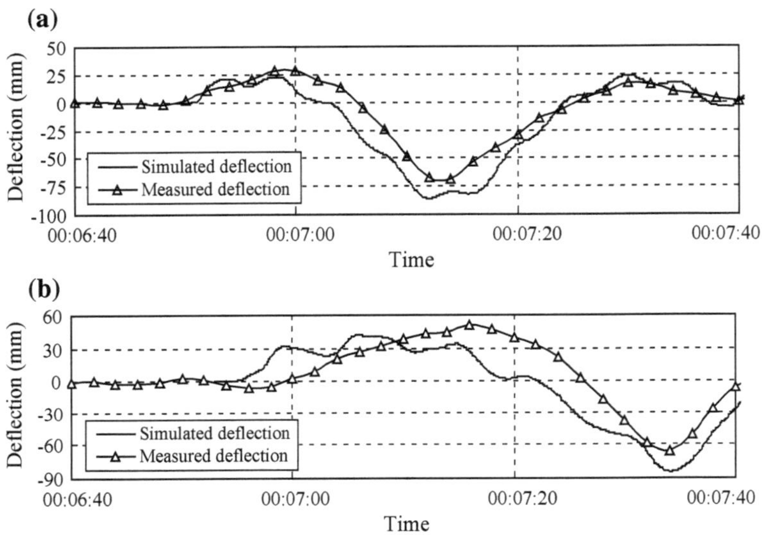

Fig. 7.15 Comparisons between finite element analysis results and uncontrolled load test data: **a** Section *l*/2 and **b** section 6*l*/8

in Fig. 7.13. The loading arrangement in finite element transient analysis of six concentrated forces is shown in Fig. 7.14. At 0:06:46, the force of Axle 6 is applied at the position of the Luzhou tower. At 0:07:35, the force of Axle 6 is applied at the position of the Yinbin tower. Then, the location of the force of Axle 6 on the finite element model can also be determined based on the speed recorded by WIM sensors. In addition, the applied position of the rest 5 concentrated forces on the finite element model depend on the axle spacing shown in Fig. 7.13.

To facilitate the application of axle loads in the finite element transient analyses, the bridge girder is meshed uniformly in the longitudinal direction, using beam4 elements. The size of elements should be carefully selected to make a tradeoff between accuracy and computational efficiency. The size of elements is set to be 0.1 m to suit the axle spacing of the Truck 3 accurately when Truck 3 was crossing the bridge deck step by step. In each step of the transient analysis, the six concentrated forces are applied on selected nodes and in the next step, each force moves to the next node. Figure 7.15 shows the measured and simulated deflections of section *l*/2 and section 6l/8. It should be noted that the measured deflections of each section in Fig. 7.15 are the averaged values of the upstream and downstream transmitters on the same section. In Fig. 7.15 the simulated time histories match generally well with the measured ones, showing the accuracy of the CPS under the actual traffic.

7.4 Deflection Assessment Using Long-Term Measurements

7.4.1 Thresholds for Deflection Assessment

The vertical deflection of main girders under the actual traffic, which can reflect the bridge's vertical stiffness, is one of the greatest concerns for bridge management staff during the service period. However, the vehicle loads caused by the actual traffic are always different from the design vehicle loads. It is necessary to develop an effective strategy to assess the actual deflection. The thresholds for the deflection assessment are presented according to the Chinese bridge design codes for the NSB. Actually, the determination of deflection thresholds should consider two levels.

The deflection threshold of the first level is the allowable deflection specified by DSHSB. This allowable deflection is considered to be the basic requirement of vertical stiffness of a suspension bridge. According to DSHSB, the maximum deflection of main girders of suspension bridges caused by the vehicle loads (excluding the impact effect) should be no more than l/300, where l is the length of main span. Based on the DSHSB, the absolute maximum value of allowable deflection is 2733 mm because the NSB's main span is 820 m.

The deflection thresholds of the second level are defined as the maximum (positive) and minimum (negative) deflections under design vehicle loads calculated by using the finite element analysis. Compared with the first level, it is obvious than the second level is a stricter requirement of vertical stiffness.

The calculation of second level includes three steps. At first, the highway class-I lane load in GCDHBC is adopted as the vehicle loads. It should be noted that the highway class-I was also utilized to design the NSB. As shown in Fig. 7.16, this lane load is composed of a concentrated force P_k and a distributed force q_k. In addition, the GCDHBC specifies two reduction factors, by which the vehicle loading effects should be multiplied after the finite element analysis. According to GCDHBC, the reduction factor of multi-lane is 0.55 on account of six lanes in the NSB and the reduction factor of bridge span is 0.94 because the main span is within the range of 800–1000 m.

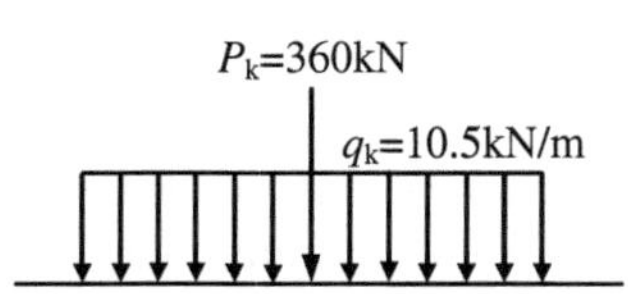

Fig. 7.16 Highway class-I lane load

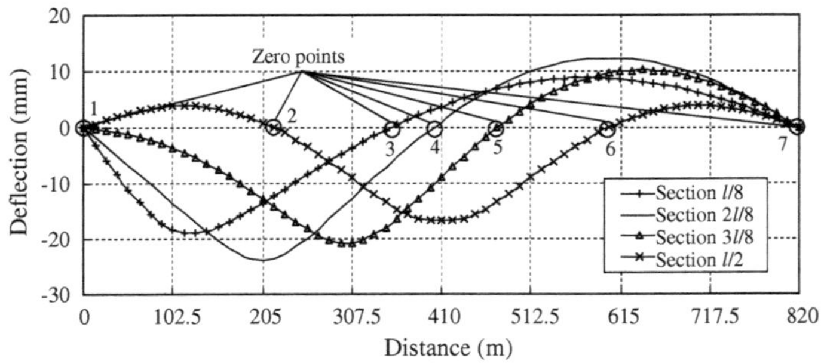

Fig. 7.17 Vertical deflection influence lines of the NSB

Table 7.2 Loading ranges in finite element analysis

Section	Deflection type	Loading ranges of distributed force q_k
$l/8$	Minimum	Zero point 1 to zero point 3
	Maximum	Zero point 3 to zero point 7
$2l/8$	Minimum	Zero point 1 to zero point 4
	Maximum	Zero point 4 to zero point 7
$3l/8$	Minimum	Zero point 1 to zero point 5
	Maximum	Zero point 5 to zero point 7
$l/2$	Minimum	Zero point 2 to zero point 6
	Maximum	Zero point 1 to zero point 2 and Zero point 6 to zero point 7

Then, the vertical deflection influence lines of the NSB are calculated by using the finite element analysis. It should be noted that the concentrated force is 100 kN in the influence line calculation. Figure 7.17 shows the deflection influence lines of sections $l/8$, $2l/8$, $3l/8$ and $l/2$. These influence lines are used to determine the loading range of the highway class-I lane load rather than calculating the maximum and minimum deflections directly because the geometrical nonlinearity needs to be considered. Table 7.2 gives the loading ranges of sections $l/8$, $2l/8$, $3l/8$, and $l/2$ to determine the maximum and minimum deflection under design vehicle loads. The loading ranges of rest sections $5l/8$, $6l/8$, and $7l/8$ can follow those of sections $3l/8$, $2l/8$, and $l/8$, respectively. After determination of loading ranges of distributed force q_k, the loading points of concentrated force P_k can also be determined according to the influence lines in Fig. 7.17.

At last, the finite element model of the NSB shown in Fig. 7.10 is used again to calculate the maximum and minimum deflections. Taking section $l/2$ as an example, according to Fig. 7.17 and Table 7.2 P_k is applied to the middle of the span and q_k is applied to the range from zero point 2 to zero point 6. Then, the minimum vertical deflection of the section $l/2$ is obtained with the value of -1352 mm. Figure 7.18 gives the results of the deflection threshold of the second level. On section $2l/8$, the maximum and minimum vertical deflections are 1242 and -1935 mm, respectively.

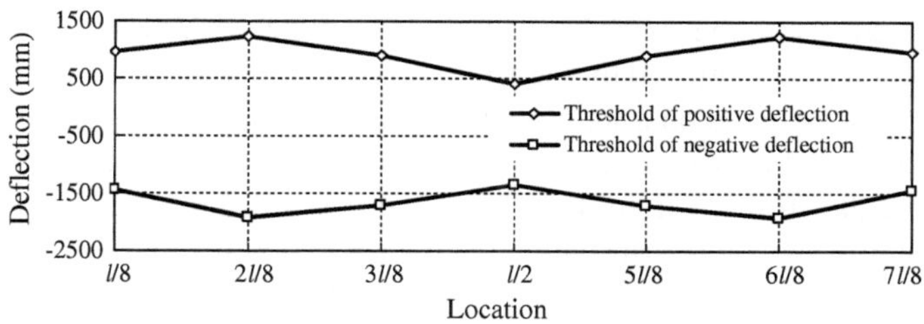

Fig. 7.18 Deflection thresholds of the second level

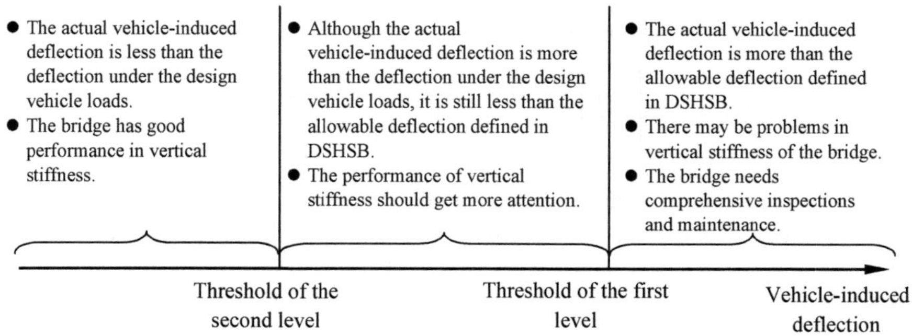

Fig. 7.19 Relationship between the first level and the second level

Obviously, the absolute values of the second level, also known as the maximum and minimum deflections under the design vehicle loads, are less than the deflection threshold of the first level, of which the value is 2733 mm from the DSHSB. In summary, the relationship between the deflection thresholds of two levels is presented in Fig. 7.19.

7.4.2 Prediction Method of Extreme Deflection

During the service time of long-span bridges, it is important to predict, as accurately as possible, the extreme deflection of main girder with a certain return period. Overestimation of the extreme deflection will lead to over maintenance costs, whereas underestimation may undermine the bridge's safety and serviceability. Using the long-term monitoring data, a reasonable estimation of the extreme deflection can be made.

EVA is a branch of statistics dealing with the extreme deviations from the median of probability distributions. EVA is widely used in civil engineering. Ni et al. [25] adopted the Gumbel distribution to predict the extreme temperatures for a certain return period using 1-year monitoring data for the Ting Kau Bridge in Hongkong. Lan et al. [26] established the extreme value distribution of gross vehicle weight of vehicle loads using the monitoring data for the Binzhou Yellow River Highway Bridge. EVA method is utilized for the estimation of extreme vertical deflection of the NSB's main girder based on a set of maxima and minima of monitored deflections. With the samples of maxima and minima from original observations, EVA attempts to parametrically fit the data to the GEVD. The GEVD unites the type I, type II, and type III extreme value distributions into a single family, to allow a continuous range of possible shapes [27, 28]. It is parameterized with a location parameter μ, a scale parameter σ, and a shape parameter, ξ. When $\xi < 0$, the GEVD is equivalent to the type III. When $\xi > 0$, the GEVD is equivalent to the type II. In the limit as ξ approaches to 0, the GEVD becomes the type I. The GEVD distribution has a cumulative distribution function as

$$F(D) = \exp\left[-\left(1 + \xi\frac{D - \mu}{\sigma}\right)^{-1/\xi}\right], 1 + \xi\frac{(D - \mu)}{\sigma} > 0, \quad (7.7)$$

where μ, σ, and ξ = constants to be determined from measurement data. The parameters are estimated by using the maximum likelihood method with 95% confidence intervals.

Given the determined parameters, rearranging Eq. (7.7) leads to

$$D = \frac{\sigma}{\xi}\left[\left(\frac{-1}{\ln F(D)}\right)^{\xi} - 1\right] + \mu \quad (7.8)$$

Denoting the sample sequence of the maxima or minima for the measured deflection by D_1, D_2, ..., D_k, where D_i's are increasingly ordered with $D_1 \leq D_2 \leq \ldots \leq D_k$ and k = number of total samples, the cumulative probability of the sample D_i is estimated as

$$p_i = \frac{i}{k + 1} \quad (7.9)$$

p_i is a rational estimate of the limiting distribution function $F(D_i)$. After the parameters determined, the quantile of p_i is estimated as

$$Q_i = \frac{\sigma}{\xi}\left[\left(\frac{-1}{\ln p_i}\right)^{\xi} - 1\right] + \mu \quad (7.10)$$

The Q versus D plot will be used to evaluate the goodness-of-fit of the estimated GEVD.

The formulated distribution model is then used to extrapolate the extremes with a certain return period. The extreme deflections D_R with a certain return period of y years are predicted by

For maximum deflection, $F(D_R) = \exp\left[-\left(1+\xi\frac{D_R-\mu}{\sigma}\right)^{-1/\xi}\right] = 1-\frac{1}{ny}$ (7.11a)

For minimum deflection, $F(D_R) = \exp\left[-\left(1+\xi\frac{D_R-\mu}{\sigma}\right)^{-1/\xi}\right] = \frac{1}{ny}$, (7.11b)

where n = extreme value number per year. When the extreme values are extracted as daily maximum or minimum values, n = 365.

7.4.3 Analysis and Assessment

Available for this study is 100 days deflection data monitored by using the CPS installed on the NSB in the first half of 2014. The original deflection data from the upstream and downstream transmitters on a certain section is averaged. To compare the extreme deflections with the thresholds in Sect. 7.4.1, the deflection component caused by actual traffic loads should be separated from the original vertical deflection data. First the wavelet transform technique is employed to eliminate the temperature effect from the averaged deflection history of a certain section. Then, the maximum and minimum values of the daily deflection history are identified. The effect of measurement noise on the maximum or minimum deflection sequence is relatively small and can be ignored. Consequently, the maximum and minimum deflection sequences can be considered to be induced by the actual traffic loads only.

EVA begins with the maximum and minimum deflection sequences with the length of 100. Table 7.3 lists the identified GEVD parameters and Fig. 7.20 and Fig. 7.21 give the Q versus D plots. It can be found that the points in the figures are sufficiently close to the best fit lines (D = Q). It means that GEVD can give an accurate fit to the maximum or minimum deflection sequences.

Table 7.3 Identified parameters of GEVD

		Section location						
		l/8	2*l*/8	3*l*/8	*l*/2	5*l*/8	6*l*/8	7*l*/8
Maximum deflection	ξ	−0.006	0.061	−0.253	−0.209	−0.142	−0.016	0.042
	σ	19.5	16.7	9.9	11.8	23	26.8	22.3
	μ	64.2	51	36.7	48.8	81.5	96.2	81.1
Minimum deflection	ξ	−0.495	−0.465	−0.473	−0.478	−0.632	−0.632	−0.551
	σ	28.6	25.4	25	24.2	38.2	48.3	32.8
	μ	−102.1	−90.6	−96.2	−110.7	−136.9	−150.7	−115.4

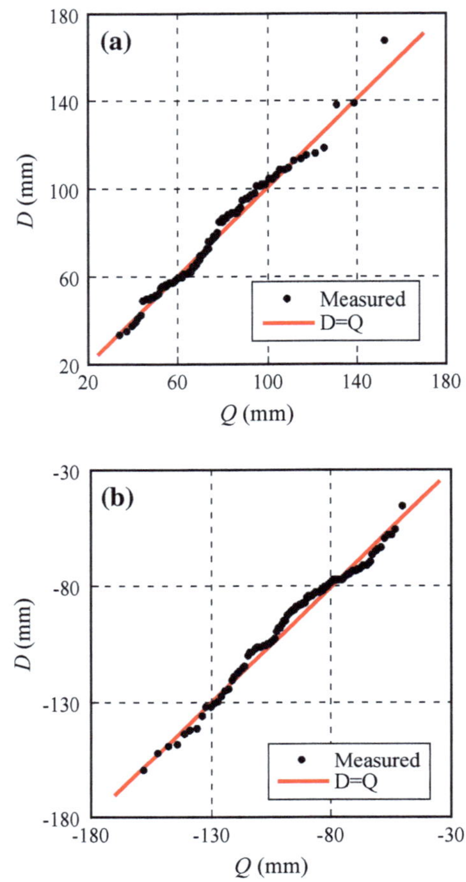

Fig. 7.20 Relations between extreme deflection *D* and quantile *Q* of section *l*/8: **a** Maximum deflection and **b** Minimum deflection

With the identified GEVD, the extreme deflections of the NSB with a return period of *y* years can be predicted. The design reference period of the NSB is 100 years and thus *y* is equal to 100. With Eq. (7.11a), the extreme positive deflection and negative deflection are obtained and shown in Fig. 7.22. For comparison, the thresholds of the second level in Fig. 7.18 are also shown in Fig. 7.22. The figure shows that the absolute values of predicted extreme deflections of all sections are less than the corresponding thresholds. It means that the design work, which is based on the design vehicle loads defined in GCDHBC, ensures enough stiffness for the NSB to bear the actual traffic loads.

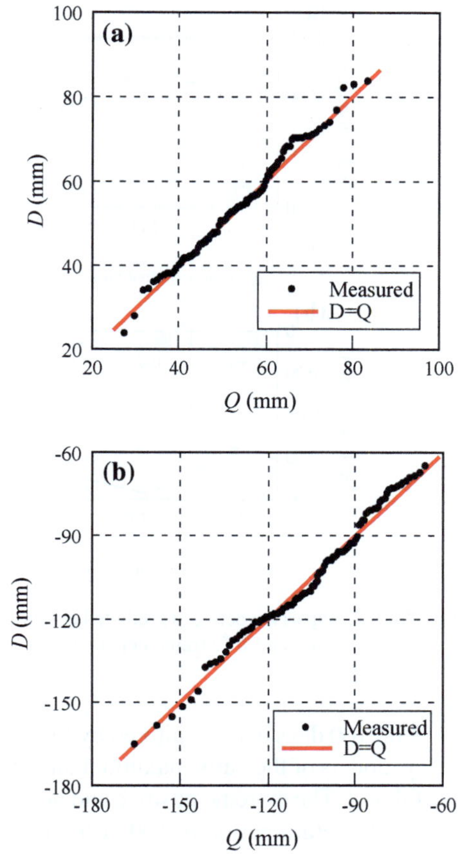

Fig. 7.21 Relations between extreme deflection D and quantile Q of section $l/2$: **a** Maximum deflection and **b** Minimum deflection

7.5 Summary

This chapter presented the CPS-based methodology of deflection monitoring and assessment for long-span bridges. With regard to tracing the profile change of the long-span bridges' main girder, the distributed pressure transmitters, which are connected with pipes, can be mounted along the main girder. The vertical deflection was measured in real time through continuously detecting the liquid pressure change in the connected pipes by pressure transmitters.

The performance of the CPS was evaluated by a series of load tests with respect to the NSB in China. In the controlled static load test, a good agreement between the measured deflections and the simulated values was observed. In the uncontrolled dynamic load test in actual traffic condition, the deflection histories induced by a heavy truck were measured and compared with the time histories from finite element analysis. The comparison showed that the CPS has an excellent capability to monitor the deflection of the suspension bridge under the actual vehicle loads.

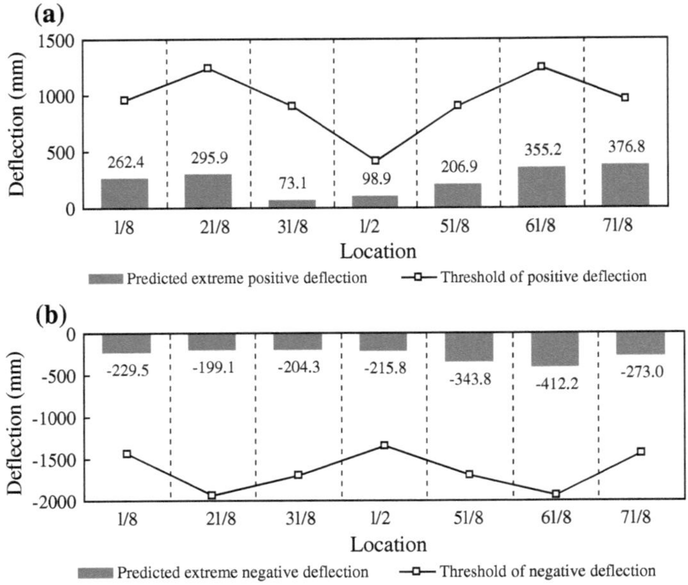

Fig. 7.22 Comparison between predicted extreme deflections and deflection thresholds: **a** Extreme positive deflection and **b** Extreme negative deflection

With 100 days data monitored by the CPS during the first half of 2014, the probability models of the daily maximum and minimum deflection were established using the GEVD. The extreme positive and negative deflections of the NSB were predicted during the return period of 100 years. For the purpose of deflection assessment, the deflection thresholds of the NSB were discussed. Comparison between the predicted extreme deflections and the thresholds was carried out. The results showed that the absolute values of the predicted extreme deflections of the main girder were less than the corresponding thresholds. The deflection monitoring and assessment methodology presented in this chapter demonstrates its excellent capability in structural health monitoring of long-span bridges.

References

1. Hou XM, Yang XS, Huang Q. Using inclinometers to measure bridge deflection. J Br Eng. 2005;10(5):564–9.
2. Ye XW, Ni YQ, Wai TT, Wong KY, Zhang XM, Xu F. A vision-based system for dynamic displacement measurement of long-span bridges: algorithm and verification. Smart Struct Syst. 2013;12(3–4):363–79.
3. Yi TH, Li HN, Gu M. Recent research and applications of GPS-based monitoring technology for high-rise structures. Struct Control Health Monit. 2013;20(5):649–70.
4. Yi TH, Li HN, Gu M. Experimental assessment of high-rate GPS receivers for deformation monitoring of bridge. Measurement. 2013;46(1):420–32.

5. Ogundipe O, Roberts GW, Brown CJ. GPS monitoring of a steel box girder viaduct. Struct Infrastruct Eng. 2014;10(1):25–40.
6. Sung YC, Miyasaka T, Lin TK, Wang CY. A case study on bridge health monitoring using position-sensitive detector technology. Struct Control Health Monit. 2012;19(2):295–308.
7. Chung W, Kim S, Kim NS, Lee HU. Deflection estimation of a full scale prestressed concrete girder using long-gauge fiber optic sensors. Constr Build Mater. 2008;22(3):394–401.
8. He X, Yang X, Zhao L. New method for high-speed railway bridge dynamic deflection measurement. J Br Eng. 2014;19(7):05014004.
9. Kohut P, Holak K, Uhl T, Ortyl L, Owerko T, Kuras P, Kocierz R. Monitoring of a civil structure's state based on noncontact measurements. Struct Health Monit Int J. 2013;12(5–6):411–29.
10. Park HS, Kim JY, Kim JG, Choi SW, Kim Y. A new position measurement system using a motion-capture camera for wind tunnel tests. Sensors. 2013;13(9):12329–44.
11. Zeng W, Yu DJ, Hu BX, Liu KM, Guo JW. An auto-monitoring system based on the communicating pipe principle of bridge deflection. J Hunan Univ (Nat Sci). 2007;34(7):44–7.
12. Zhu Y, Fu Y, Chen W, Huang S. Online deflection monitoring system for Dafosi cable-stayed bridge. J Intell Mater Syst Struct. 2006;17(8–9):701–7.
13. Yang JC, Chen WM. Connected pipe opto-electronic deflection measurement system and the application in Dafoshi Yangtze river bridge. J Optoelectron Laser. 2006;17(3):343–6.
14. Zhu SF, Zhou ZX, Wu HJ. Application of semi-closed connected pipe differential pressure sensor in bridge deflection measurement. Transducer Microsyst Technol. 2014;33(1):150–3.
15. Meng X, Dodson AH, Roberts GW. Detecting bridge dynamics with GPS and triaxial accelerometers. Eng Struct. 2007;29(11):3178–84.
16. Kaloop MR. Bridge safety monitoring based-GPS technique: case study Zhujiang Huangpu Bridge. Smart Struct Syst. 2012;9(6):473–87.
17. Psimoulis PA, Stiros SC. Measurement of deflections and of oscillation frequencies of engineering structures using robotic theodolites (RTS). Eng Struct. 2007;29(12):3312–24.
18. Gentile C, Bernardini G. An interferometric radar for non-contact measurement of deflections on civil engineering structures: laboratory and full-scale tests. Struct Infrastruct Eng. 2010;6(5):521–34.
19. Gentile C. Deflection measurement on vibrating stay cables by non-contact microwave interferometer. NDT&E Int. 2010;43(3):231–40.
20. Ministry of Transport of the People's Republic of China. Design specification for highway suspension bridge. Beijing: China Communications Press; 2002.
21. Ministry of Transport of the People's Republic of China. General code for design highway bridges and culverts JTG D60-2004. Beijing: China Communications Press; 2004.
22. Lan Z, Yang X, Chen W, Zhou J, Zhou Z, Huang Z, Zhang B. Study on non-contact weighted-stretched-wire system for measuring bridge deflections and its effect factors. Eng Struct. 2008;30(9):2413–9.
23. Cao Y, Yim J, Zhao Y, Wang ML. Temperature effects on cable stayed bridge using health monitoring system: a case study. Struct Health Monit-An Int J. 2010;10(5):523–37.
24. Reda TMM, Noureldin A, Lucero JL, Baca TJ. Wavelet transform for structural health monitoring: a compendium of uses and features. Struct Health Monit-An Int J. 2006;5(3):267–95.
25. Ni YQ, Hua XG, Wong KY, Ko JM. Assessment of bridge expansion joints using long-term displacement and temperature measurement. J Perform Constr Facil. 2007;21(2):143–51.
26. Lan C, Li H, Ou J. Traffic load modelling based on structural health monitoring data. Struct Infrastruct Eng. 2011;7(5):379–86.
27. Castillo E. Extreme value theory in engineering. New York: Academic Press; 1988.
28. Coles S. An introduction to statistical modeling of extreme values. London: Springer; 2001.

Chapter 8
Serviceability Assessment Based on Deflection Measurements

8.1 Introduction

Service load deformations may cause deterioration of wearing surfaces and local cracking in concrete slabs and in metal bridges that could impair serviceability and durability, even if self-limiting and not a potential source of collapse. In most recent AASHTO specifications, although the criteria for deflection control are made optional at the discretion of bridge owners or designers, the deflection limitation provisions shall be considered mandatory for orthotropic decks, lightweight decks comprised of metal and concrete, such as filled and partially filled grid decks, and unfilled grid decks composite with reinforced concrete slabs [1]. In the same manner, the latest Chinese specifications for design of highway suspension bridge adopt a new provision, which is that the traffic-induced vertical deflection of suspension bridge's main girder should be less than 1/250 of the main span [2]. In addition, the Chinese guidelines for design of highway cable-stayed bridge require that the traffic-induced vertical deflection of steel girder should be less than 1/400 of the main span [3]. For a new bridge design, the live load deflection is often predicted by applying the design traffic loads on the bridge analytical model and the predicted deflection is then compared with the design limits provided by the design specifications [4, 5]. However, calculated bridge deflections have often been found to be difficult to verify in the field due to numerous sources of stiffness not accounted for in calculations [1].

With the rapid developments of structural health monitoring system (SHMS) techniques, the deflection caused by live load for in-service bridges can be studied based on high-quality monitoring data from advanced deflection sensing systems. During the past years, various deflection sensing systems have been developed and a large amount of monitoring data of bridge deflection become available, such as GPS, noncontact vision sensors, interferometric radar systems, connected pipe systems, etc. [6–9]. Similar to other collected data from SHMS, a major challenge still remains in terms of how to effectively utilize the monitoring data of bridge deflection under live loads to improve the knowledge about the bridge performance and design.

Y. Deng and A. Li, *Structural Health Monitoring for Suspension Bridges*,
https://doi.org/10.1007/978-981-13-3347-7_8

Extreme value analyses and applications of reliability theory have been conducted by several researchers with on-site load test data or in-service monitored data to evaluate the performance of bridges, primarily focusing on structural strength performance or ultimate limit state [10–15]. For instance, Bhattacharya et al. [11] used site-specific strain measurements for bridge rating at two different limit states, yield and plastic collapse. In their study, the Bayesian updating of the empirical distribution function was employed to obtain an extreme value distribution of strain caused by live loads. Messervey et al. [13] presented a study of conducting the reliability assessment based on the 90-day monitoring data to obtain the extreme value distributions of the in-service strain. Gokce et al. [15] applied the system reliability theory in the structural strength assessment for a multi-girder movable bridge with the help of the monitored information from six strain sensors.

Despite the reported efforts on investigating the strength performance of bridges with SHM data as summarized above, existing studies about the serviceability limit state, such as bridge deflection performance under live loads, are rare. Besides, there are still some technical challenges associated with utilizing the deflection monitoring data that need to be addressed. First, an accurate and continuous deflection monitoring technique for long-span bridges needs to be carefully selected to provide reliable monitoring data [14]. In addition, how to develop an appropriate reliability model with monitored deflection data from multiple sensors also deserves further investigation [7]. This study presents a deflection assessment framework of long-span bridges based on the long-term monitoring data from an advanced deflection monitoring system. Theories of extreme value and system reliability are applied to develop such a framework, after the serviceability limit state with specified service times and target reliability are appropriately set up. The proposed framework is applied to a prototype suspension bridge to conduct the in-service traffic deflection reliability assessment.

8.2 Serviceability Assessment Framework

For long-span bridges, the effect of dead load dominates the total vertical deflection while the magnitude of deflection induced by live loads is not significant. The dead load induced deflection takes place step by step along with the process of bridge construction. During construction process, the dead load induced deflection is often counteracted by means of construction control to obtain the expected profile. After construction completion, traffic loads begin to influence bridge performance.

The proposed framework has five major steps as depicted in Fig. 8.1: (1) the main purpose of the first step is local extremes extracting of traffic-induced deflection from the raw data collected by the SHMS. In this study, the deflection caused by temperature is first eliminated from the original deflection monitoring data and the local extremes of the deflection caused by moving traffic are extracted; (2) The second step is to present a probabilistic model for the extracted local extremes of the traffic-induced deflection. The POT approach is employed to obtain the exceedances of the local extremes. Then, the GPD is fitted to the exceedances of the local extremes

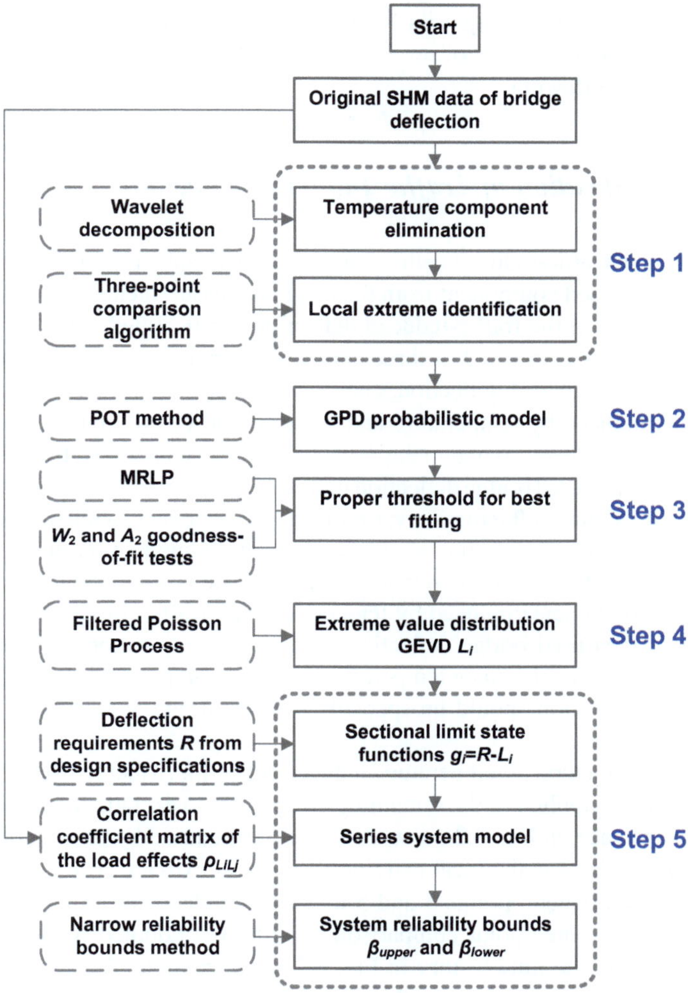

Fig. 8.1 SHM-based assessment framework of vertical deflection for long-span bridges

to obtain the exceedance distribution; (3) The threshold u is a key parameter of the GPD-based exceedance distribution. Hence, a new strategy of finding the proper threshold of the GPD is developed based on the mean residual life plot (MRLP) and the goodness-of-fit tests in the third step; (4) The fourth step is to theoretically extend the GPD-based exceedance distribution to the extreme value distribution. It is found that the extreme value distribution yields to the GEVD; (5) The last step is aiming to establish the reliability model of the traffic-induced deflection for long-span bridges. The serviceability limit state of the traffic-induced deflection for each instrumented bridge section is established with the deflection limit and the extreme value distribution. Finally, a quasi-system reliability assessment of the whole bridge is made based on the concept of series system reliability. More details are given for each step in the following.

8.2.1 Identification of Traffic-Induced Deflection

Identification of traffic-induced deflection includes two sub-steps: (1) elimination of temperature-induced component from the original measurements and (2) extraction of local extremes of the traffic-induced deflection. Vertical deflection of long-span bridges is typically composed of three components including temperature-induced deflection, traffic-induced deflection, and noise-induced pseudo deflection [7]. The components induced by temperature and noise should be eliminated from the raw deflection measurements to obtain the deflection component caused by traffic. Existing studies on measured bridge deflection indicated that temperature-induced component is usually a diurnal fluctuation variation at low frequency, while traffic-induced component is a transient fibrillation at a relatively high range of frequency. In terms of noise-induced deflection component, the magnitude is significantly smaller than those of the components induced by temperature and traffic. In view of such characteristics of measured bridge deflection, wavelet decomposition is recommended as an effective tool to eliminate temperature-induced component [16, 17]. Note that wavelet decomposition should be specially structured and adjusted for a specific bridge.

After the elimination of temperature-induced deflection, a local extreme searching algorithm will be applied to the remaining component to exclude the non-inflection points. For instance, in Fig. 8.2 points 3, 4, and 5 are caused by passing of the same vehicle while point 4 is the local extreme. If no local extreme searching is applied to the traffic component, points 3 and 5 may be included even if the POT method is used. Hence, a three-point comparison algorithm is adopted for local extreme searching. Taking the time history in Fig. 8.2 as an example, the algorithm is that points S_2, S_3 and S_4 are compared, thus S_3 is deleted because of $S_2 > S_3 > S_4$. Then, the comparison of points S_2, S_4 and S_5 leads to a fact that S_4 is a valley local extreme because of $S_4 < S_2$ and $S_4 < S_5$. The peak local extremes can be also extracted by using a similar procedure. The flowchart of traffic-induced deflection identification is shown in Fig. 8.2.

8.2.2 POT-Based Probabilistic Model

Currently, many approaches have been developed to fit a distribution to the tail of the cumulative distribution function (CDF) of traffic-induced effects [18–21]. This is mainly since the vehicle weights, which can induce bridge deflection, usually vary in a wide range. Besides, heavier vehicles tend to dominate the traffic load on a bridge, with the lighter vehicles making very little contribution to the probability of

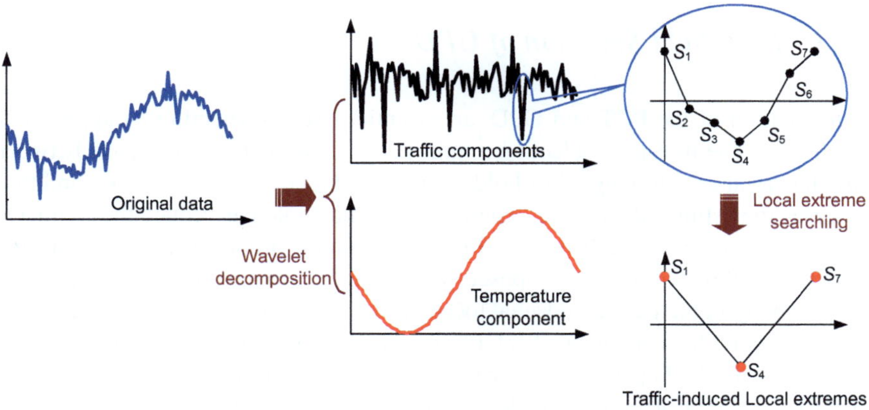

Fig. 8.2 Flowchart of identification of traffic-induced deflection

exceedance of the extremes. Two particularly popular tail fitting approaches are the block maximum [7, 13, 22] and the POT [20, 23]. The disadvantage of the block maximum approach is that only the maximum or minimum values of live load effect in each time block are considered, thus some useful information may be lost. Hence, the POT approach is adopted in the present study to fit the GPD to the tail of the monitored vertical deflection data. The POT-based modeling is briefly described as follows:

X_1, X_2, ..., X_n are a series of independent and identically distributed random variables. The threshold is defined as a selected value u. If $X_i > u$, X_i is called an exceedance. Accordingly, the distribution function of exceedances $F_u(x)$ is

$$F_u(x) = P(X \le x | X > u) = \frac{F(x) - F(u)}{1 - F(u)}, \quad x \ge u \tag{8.1}$$

The probability density function (PDF) $f_u(x)$ is

$$f_u(x) = \frac{f(x)}{1 - F(u)}, \quad x \ge u \tag{8.2}$$

The exceedance distribution function can be deduced precisely from a given distribution function $F(x)$. The GPD is the asymptotic distribution for the exceedances when $F(x)$ is unknown. Balkema and Haan [24] and Pickands [25] proposed that the exceedance distribution approximately yields to GPD when threshold u is sufficiently large. The CDF of GPD is defined as

$$F_u(x; u, \sigma, \xi) = 1 - \left(1 + \xi \frac{x - u}{\sigma}\right)^{-1/\xi}, \quad x \ge u, \quad 1 + \xi \frac{x - u}{\sigma} > 0 \tag{8.3}$$

where u is the selected threshold, and ξ and σ are the shape and scale parameters, respectively.

8.2.3 Threshold Selection of GPD

In application of the POT and GPD, an essential issue is selecting an appropriate threshold u. The threshold selection requires consideration of the trade-off between bias and variance: a too high threshold reduces the number of exceedances, and thus increases the estimated variance, whereas a low threshold can reduce the estimated variance but increase the bias [26]. In order to find a proper threshold u, several statistical methods have been proposed, which can be grouped into three categories [27]: (1) nonparametric methods that locate the changing point between extreme and non-extreme regions of the data [27]; (2) graphical methods that search for the linear behavior of the GPD parameters (or related metrics) with the increasing threshold [20]; (3) goodness-of-fit tests that locate the lowest threshold u that a GPD model is applicable for any given level of significance [28].

A threshold selection strategy is developed in this study with the joint usage of a graphical method and the goodness-of-fit tests. The most popular graphical method is based on the mean residual life plot (MRLP). In an MRLP, a mean excess function (MEF) is defined as

$$e(u) = E(X - u|X > u) = \frac{\sigma + \xi u}{1 - \xi} \tag{8.4}$$

where $e(u)$ represents the MEF and $E(\cdot)$ is the expectation operation. Based on Eq. (8.4), a proper threshold can be determined by (1) plotting $e(u)$ as a function of u, and (2) identifying the lowest u above which the relationship between $e(u)$ and u begins to be linear.

However, MRLP requires the practitioner to have substantial expertise and can be rather subjective [29]. Choulakian and Stephens [28] suggested the use of goodness-of-fit tests to identify the lowest threshold above which the exceedances can be approximated by a GPD model. Referred to as "failure-to-reject", the threshold increases until the null hypothesis (H_0) of GPD is not rejected at a desired significance level. Choulakian and Stephens [28] proposed A^2 Anderson–Darling and W^2 Crámer–von Mises tests for GPD:

$$W^2 = \frac{1}{12n} + \sum_{i=1}^{n}\left[F(X_i) - \frac{2i-1}{2n}\right] \tag{8.5}$$

$$A^2 = -n - \sum_{i=1}^{n}\frac{2i-1}{n}\left\{\log[F(X_i)] + \log\left[1 - F(X_{n+1-i})\right]\right\} \tag{8.6}$$

where $X_1 \leq X_2 \leq \cdots \leq X_n$ are ascending order statistics, and F is the theoretical CDF tested for fitting. Both A^2 and W^2 quantify the deviations between a selected theoretical distribution, F, and the empirical CDF.

In engineering practice, the MRLP method may have another limitation besides the subjective judgment. The lowest threshold, above which the MEF versus u is linear, may not allow the best fitting of GPD for the exceedances. In terms of the goodness-of-fit tests, the lowest threshold, above which the exceedances can be approximated by a GPD model, may also not yield to the best fitting of GPD. It should be reminded that the fundamental purpose of the threshold selection of GPD is to pursue a best fitting for the exceedances. Hence, the proposed strategy is that when a linear relationship between the MEF and u is observed, the threshold that can result in the best fitting of the GPD is the proper one.

8.2.4 Extreme Value Distribution

Filtered Poisson process has been adopted as the probabilistic model for simulating vehicle loads [30]. Considering the similarities between vehicle loads and vehicle-induced deflection, filtered Poisson process is also adopted as the probabilistic model for vehicle-induced deflection in this study. As shown in Fig. 8.3, the vehicle-induced effects are composed of a series of random pulses.

In Fig. 8.3, the occurrence time and duration of vehicle-induced pulse S_i are T_i and τ_i, respectively. Generally, the duration is relatively short as compared to the bridge service time. When S_i (i=1, 2, ..., k) are independently and identically distributed, the probability of the occurrence number $N(T)$ of the load effects during the time interval (0, T) is [31]:

$$P[N(T) = k] = \frac{(\lambda T)^k \exp(-\lambda T)}{k!}, \quad k = 0, 1, 2, \ldots \tag{8.7}$$

where λ is the incidence rate. The extreme value distribution of vehicle-induced deflection is

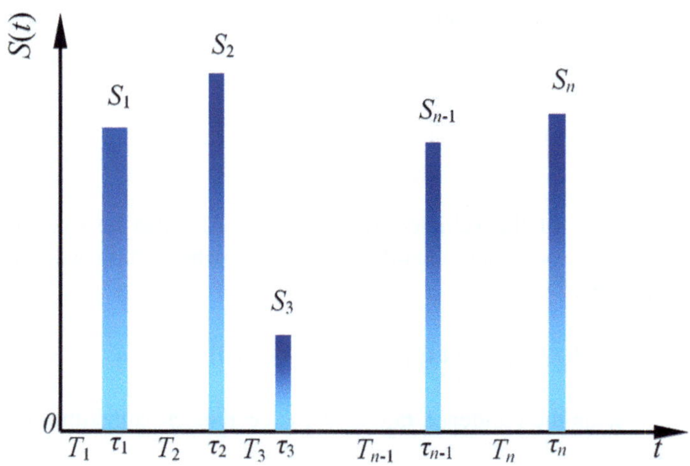

Fig. 8.3 Schematic diagram of filtered Poisson process

$$
\begin{aligned}
F_{S_{\max}}(s) = P(S_{\max} \le s) &= \sum_{k=0}^{\infty} P[S_{\max} \le s | N(T) = k] \cdot P[N(T) = k] \\
&= \sum_{k=0}^{\infty} P[S_1 \le s, \ldots, S_n \le s | N(T) = k] \cdot P[N(T) = k]
\end{aligned}
\tag{8.8}
$$

Because S_i $(i=1, 2, \ldots, k)$ are independent and identically distributed, $F_{S_{\max}}(s)$ can be rewritten as

$$
F_{S_{\max}}(s) = \sum_{k=0}^{\infty} [P(S_i \le s)]^k \cdot P[N(T) = k] \tag{8.9}
$$

With the given CDF of the vehicle-induced effects $F(s)$, substituting Eq. (8.7) into Eq. (8.9) leads to

$$
\begin{aligned}
F_{S_{\max}}(s) &= \sum_{k=0}^{\infty} [F(s)]^k \cdot \frac{(\lambda T)^k \exp(-\lambda T)}{k!} = \exp(-\lambda T) \sum_{k=0}^{\infty} \frac{[F(s)\lambda T]^k}{k!} \\
&= \exp(-\lambda T) \cdot \exp[F(s)\lambda T] = \exp\{-\lambda T[1 - F(s)]\}
\end{aligned}
\tag{8.10}
$$

As shown in Eq. (8.10), several factors influence the extreme value distribution of vehicle-induced deflection, such as incidence rate λ, service time T and probabilistic distribution of vehicle-induced effect $F(s)$.

The distribution function of exceedance has been established based on the GPD in Step 2. Generally, the extreme value distribution $F_{S_{\max}}(s)$ is significantly affected by the right tail shape of $F(s)$ [26]. Hence, the GPD-based distribution function of exceedance is recommended to obtain the extreme value distribution $F_{S_{\max}}(s)$.

Letting $F_u(s)$ be the distribution function of exceedance of $F(s)$ with the threshold u, Eq. (8.1) leads to

$$
F_u(s) = \begin{cases} \frac{F(s)-F(u)}{1-F(u)} & s \ge u \\ 0 & s < u \end{cases} \tag{8.11}
$$

The exceedances of the vehicle-induced effects can also be simulated by the filtered Poisson process. Hence, according to Eq. (8.10) the extreme value distribution of the exceedances is

$$
F_{u,S_{\max}}(s) = \exp\{-\lambda_u T[1 - F_u(s)]\}, \quad s \ge u. \tag{8.12}
$$

where λ_u represents the incidence rate of the exceedances. Substituting Eq. (8.11) into Eq. (8.12) results in

$$F_{u,S_{\max}}(s) = \exp\left\{-\frac{\lambda_u T}{1 - F(u)}[1 - F(s)]\right\}, \quad s \geq u \tag{8.13}$$

During the same time interval $(0, T]$, the incidence rate of the exceedances λ_u can be obtained approximately from the incidence rate of the original vehicle-induced effects λ

$$\lambda_u = \lambda[1 - F(u)] \tag{8.14}$$

Substituting Eq. (8.14) into Eq. (8.13) leads to a new expression of the extreme value distribution of the exceedances

$$F_{u,S_{\max}}(s) = \exp\{-\lambda T[1 - F(s)]\} = F_{S_{\max}}(s) \tag{8.15}$$

Equation (8.15) suggests that the extreme value distribution of the exceedances $F_{u,S_{\max}}(s)$ is in agreement with the extreme value distribution of vehicle-induced deflection $F_{S_{\max}}(s)$. Hence, $F_u(s)$ provides an alternative of $F(s)$ to obtain $F_{S_{\max}}(s)$. As shown in Eq. (8.3), the GPD was adopted as the CDF of the exceedances in Step 2. Changing x into s in Eq. (8.3) and then substituting Eq. (8.3) into Eq. (8.12) leads to the extreme value distribution of vehicle-induced effects $F_{S_{\max}}(s)$

$$F_{S_{\max}}(s) = F_{u,S_{\max}}(s) = \exp\left[-\lambda_u T\left(1 + \xi\frac{s - u}{\sigma}\right)^{-1/\xi}\right] \tag{8.16}$$

where T represents the bridge service time. Letting

$$\tilde{u} = u - \frac{\sigma}{\xi}\left[1 - (\lambda_u T)^{\xi}\right], \quad \tilde{\sigma} = \sigma(\lambda_u T)^{\xi}, \quad \tilde{\xi} = \xi \tag{8.17}$$

One can rewrite Eq. (8.16) as [32, 33]

$$F_{S_{\max}}(s) = \exp\left[-\left(1 + \tilde{\xi}\frac{s - \tilde{u}}{\tilde{\sigma}}\right)^{-1/\tilde{\xi}}\right] \tag{8.18}$$

Equation (8.18) implies that when GPD is adopted as the CDF of the exceedances, the corresponding extreme value distribution yields to the GEVD.

8.2.5 Quasi-System Reliability Assessment for Vertical Deflections

The reliability index defined as the probabilistic indicator of a given limit state is adopted to assess the performance of the bridge vertical deflection induced by traffic in this study. With the extreme value distributions of the vertical deflections for several monitoring sections on a long-span bridge, a multi-sensor-based system reliability assessment approach is developed. Suppose that there are m sections instrumented with deflection sensors, the section reliability of the vertical deflection can be obtained by using the following limit state function:

$$g_i = R_i - L_i = 0 \quad (i = 1, 2, 3, \ldots, m) \tag{8.19}$$

$$\beta_i = \Phi^{-1}\left(1 - P_{f_i}\right) \tag{8.20}$$

where g_i is the limit state function of the ith section; β_i is the reliability index of the ith section; m is the number of the equipped sections in the bridges; $\Phi^{-1}(\cdot)$ is the inverse CDF of standard normal distribution. R_i is the deflection limit which is related to the serviceability requirement. R_i can be the allowable deflection specified by the bridge design codes or the calculated deflection under the design vehicle loads. The discussion on the deflection limit will be presented in the following section. Note that the deflection limit R_i is considered to a deterministic constant and L_i is the extreme traffic-induced vertical deflection of the ith section. The system model of the bridge is defined based on the assumption of the weakest link. In other words, in a series system in which any section vertical deflection can fail to meet the requirement, the system is defined as the most critical failure mode. Each instrumented section is simulated as a member in the series system. In fact, the series system described in this study is different from the usual sense of structural series system. The concept of structural series system is borrowed to model the relation between the reliabilities of the instrumented sections and the overall probabilistic performance of the main girder. Hence, the reliability model of the in-service deflection can be called a quasi-series system model. It should be noted that the system failure defined in this study corresponds to an alert for the bridge serviceability condition deserving further attention because of the excessive vertical deflection, rather than a structural failure such as collapses in a traditional sense.

Ditlevsen [34] developed a method of narrow reliability bounds, accounting for the actual correlation between the members of the series system. The narrow bounds can be written as

$$P_{f,\text{lower}} \leq P_f \leq P_{f,\text{upper}}$$

$$P_{f,\text{upper}} = \sum_{i=1}^{m} P_{f_i} - \sum_{i=2,j<i}^{m} \max\left(P_{f_{ij}}\right)$$

$$P_{f,\text{lower}} = P_{f_1} + \sum_{i=2}^{m} \max\left[P_{f_i} - \sum_{j=1}^{i-1} P_{f_{ij}}, 0\right] \tag{8.21}$$

where P_f is the failure probability of the system; P_{f_i} is the failure probability of the ith mode (section) of which the vertical deflection exceeds the predefined requirement.

$P_{f_{ij}}$, which can be computed by numerical integration, is the joint failure probability of the ith and jth modes (sections). The best bounds are obtained by rank ordering the failure events from the highest probability of failure to the lowest [34]. So P_{f_1} represents the highest failure probability. The joint probability $P_{f_{ij}}$ is computer by numerically integrating the bivariate normal distribution function [35]:

$$P_{f_{ij}} = \int_{\beta_i}^{\infty}\int_{\beta_j}^{\infty} \frac{1}{2\pi\sqrt{1-\rho_{ij}^2}} \exp\left[-\frac{1}{1\left(1-\rho_{ij}^2\right)}\left(x_i^2 + x_j^2 - 2\rho_{ij}x_i x_j\right)\right] \mathrm{d}x_i \mathrm{d}x_j \quad (8.22)$$

where β_i and β_j are the reliability indices for the ith and jth modes, respectively. ρ_{ij} represents the correlation coefficient between the ith and jth modes.

With the limit state function shown in Eq. (8.19), it can be easily found that ρ_{ij} only depends on the random variables L_i and L_j because the resistance of the bridge vertical deflection is a constant in this study. ρ_{ij} can be computed as

$$\begin{aligned} \rho_{ij} &= \frac{E\{[g_i - E(g_i)]\cdot[g_j - E(g_j)]\}}{\sqrt{\left[E(g_i^2) - E^2(g_i)\right]\cdot\left[E\left(g_j^2\right) - E^2(g_j)\right]}} \\ &= \frac{E\{[R - L_i - E(R - L_i)]\cdot[R - L_j - E(R - L_j)]\}}{\sqrt{\left\{E\left[(R - L_i)^2\right] - E^2(R - L_i)\right\}\cdot\left\{E\left[(R - L_j)^2\right] - E^2(R - L_j)\right\}}} \\ &= \frac{E\{[L_i - E(L_i)]\cdot[L_j - E(L_j)]\}}{\sqrt{\left[E(L_i^2) - E^2(L_i)\right]\cdot\left[E\left(L_j^2\right) - E^2(L_j)\right]}} = \rho_{L_iL_j} \end{aligned} \quad (8.23)$$

where $\rho_{L_iL_j}$, which can be obtained by using the SHM data, represents the correlation coefficient of the random variables L_i and L_j. With the estimated bounds of the failure probability, the reliability bounds can be determined as

$$\beta_{\text{upper}} = \Phi^{-1}\left(1 - P_{f,\text{lower}}\right)$$

$$\beta_{\text{lower}} = \Phi^{-1}\left(1 - P_{f,\text{upper}}\right) \quad (8.24)$$

8.3 Application of the Proposed Framework

8.3.1 Measurement Processing

The prototype bridge of this chapter is also the Nanxi Suspension Bridge (NSB). The CPS-based deflection monitoring system has been introduced in Chap. 7 in details.

Table 8.1 Trucks measured by the WIM system on February 14, 2014

Truck number	Time	Descriptions of vehicle types	Lane location	Speed (km/h)	GVW (kN)
1	00:06:46	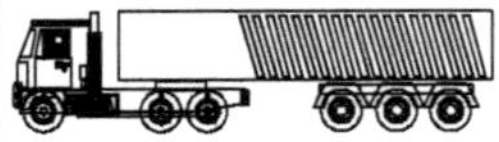	4	61	503
2	00:08:35		1	56	167
3	00:20:11		4	59	390
4	00:22:28		4	73	199
5	00:23:25		1	80	232
6	00:27:56		4	63	561

Processing the original monitoring data includes identification of traffic-induced deflection and correlation analysis of monitored vertical deflection data. First, the data of sensor S5 (section 3 shown in Fig. 7.2) collected on February 14, 2014 as shown in Fig. 8.4a is taken as an example to demonstrate the step of identification of traffic-induced deflection. The wavelet decomposition with Daubechies 25 is used to decompose the original monitored data at level 12. The temperature-induce component is extracted from the monitored deflection data and shown in Fig. 8.4b. The remaining data excluding the temperature-induced component is shown in Fig. 8.4c. Figure 8.4d shows a zoomed-in view of the deflection without temperature effects measured from 00:00 to 00:30. Six truck events can be identified in this figure. These truck events can be found in the traffic data measured by the WIM system. From 0:00 to 00:30, there were 53 vehicles recorded by the WIM system. Table 8.1 lists the truck information including the arriving time, vehicle type, lane location, speed, and gross vehicle weight (GVW). All rest 47 vehicles were passenger cars or mini trucks with GVWs less than 100 kN. In this table, the arriving time means a vehicle arrived at the location of the WIM sensor. Linking the arriving times in Table 8.1 and the truck events in Fig. 8.4d indicates that the deflection data measured by the CPS and the traffic information by the WIM system is consistent in time. In addition, Fig. 8.4e presents the relationship between the GVWs and the maximum deflections induced by these trucks events. D_{max}, which is identified from the curve in Fig. 8.4d, is the maximum deflection of each truck events. The scatters in this figure distribute around the regression line. The relationship in this figure reveals that the vertical deflection measured by the CPS can accurately capture the features of the in situ traffic loads.

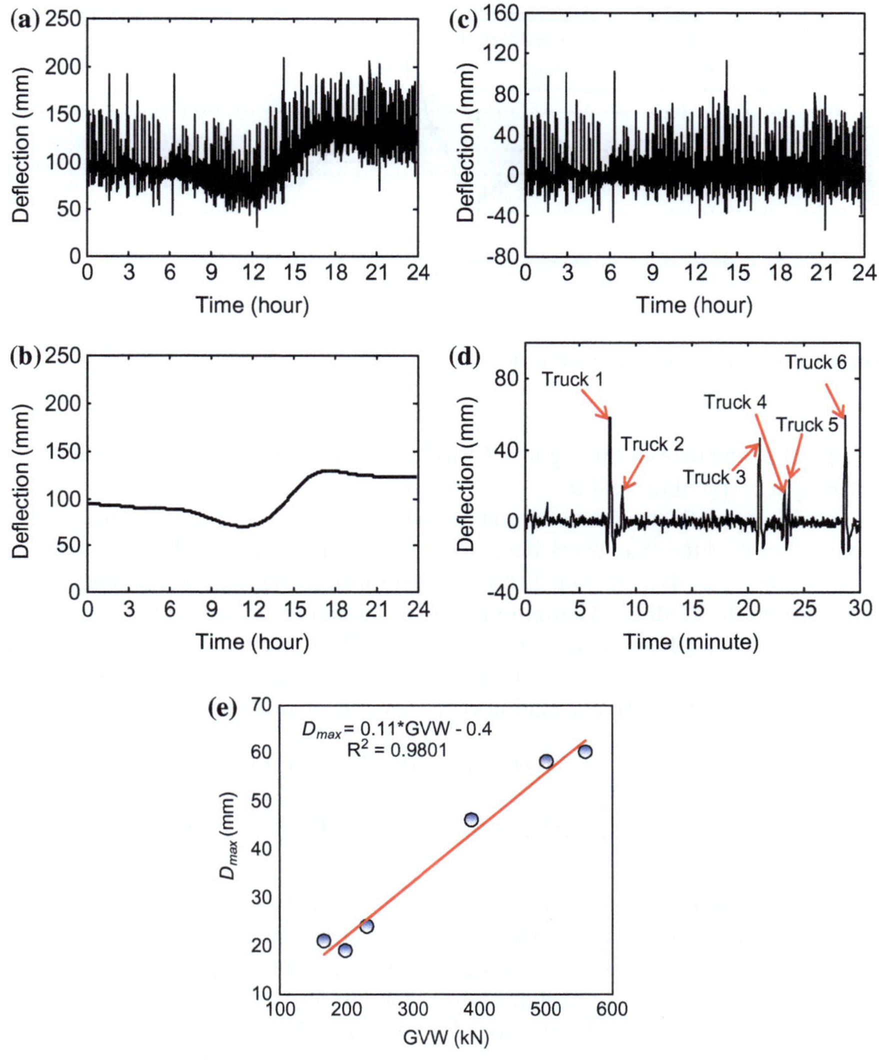

Fig. 8.4 Monitoring data of February 14, 2014: **a** original data, **b** temperature-induced deflection, **c** deflection without temperature effects, **d** traffic and noise components and **e** relationship between the GVW and the maximum deflections of the truck events

Figure 8.5 presents a typical vehicle event between time 00:20 to 00:22 and the local extreme points are extracted by using the three-point comparison algorithm. In this figure, the local extreme points are divided into two parts. The first part marked with dash lines, which could be called pseudo extreme points, is almost in the range of −5 to 5 mm. The first part can be ascribed to the testing noise or some other interferences. The second part contains three points, which are considered to be the authentic extreme points caused by the vehicle loads. When the POT method

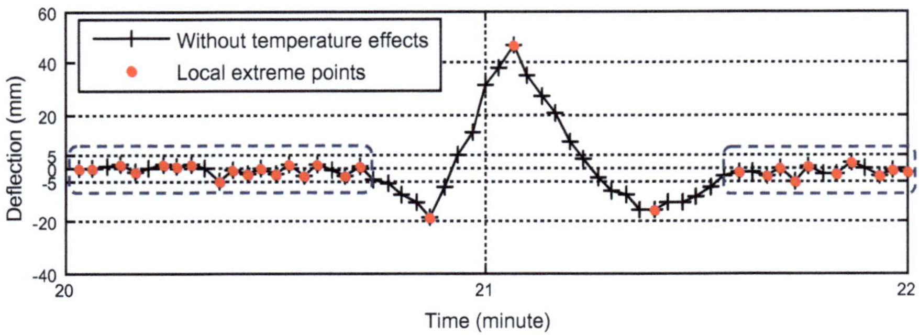

Fig. 8.5 Extraction of local extreme points

is applied to the data in this figure, the pseudo extreme points can be excluded by choosing a proper threshold u.

The 25-day deflection data measured during February of 2014 are studied as a demonstration of the framework developed in this study. The sampling frequency of all 15 sensors is 0.5 Hz, and thus 1,080,000 records of each sensor were collected and saved during the 25 days. After eliminating the components caused by temperature from the original deflection data, the remaining deflection data from a certain sensor is expressed as D_i ($i = 1, 2, \ldots, 14$). Note that the noise component is still included in D_i ($i = 1, 2, \ldots, 14$). The correlation coefficients $\rho_{D_iD_j}$ of D_i and D_j is computed as

$$\rho_{D_iD_j} = \frac{E\{[D_i - E(D_i)] \cdot [D_j - E(D_j)]\}}{\sqrt{[E(D_i^2) - E^2(D_i)] \cdot [E(D_j^2) - E^2(D_j)]}} \tag{8.25}$$

The correlation coefficients of the vertical deflection of the 14 sensors are shown in Fig. 8.6. The correlation coefficients between the upstream and downstream sensors are positive and larger than 0.9. Taking sensor S5 as an example, Fig. 8.6b shows the values of $\rho_{D_5D_j}$. To help explain the distribution of $\rho_{D_5D_j}$, Fig. 8.6c presents the influence lines of the vertical deflection. The vertical deflection influence lines (under 100 kN concentrated force) of the concerned sections, developed by the authors in the previous study [7], are presented in Fig. 8.6c. First, the correlation coefficients $\rho_{D_5D_j}$ are found to be positive or negative. For example, $\rho_{D_5D_3}$ and $\rho_{D_5D_9}$ are 0.30 and −0.06, respectively. D_5, D_3, and D_9 belong to sections 3, 2, and 5, respectively. In Fig. 8.6c, the influence lines indicate that when the concentrated force is applied to section 3, the deflections of sections 3, 2, and 5 are −20.8, −12.7, and 4.1 mm, respectively. Hence, D_5 and D_3 are positively correlated while D_5 and D_9 are negatively correlated.

Second, $\rho_{D_5D_9}$ and $\rho_{D_5D_{14}}$ are −0.06 and −0.47, respectively. These values imply that the negative correlation between sections 3 and 7 are found to be stronger than that between sections 3 and 5. Note that $D14$ belongs to section 7. However, the spatial distance between sections 3 and 5 are closer than that between sections 3 and 7. This can be explained also by the influence lines. Applying the concentrated force

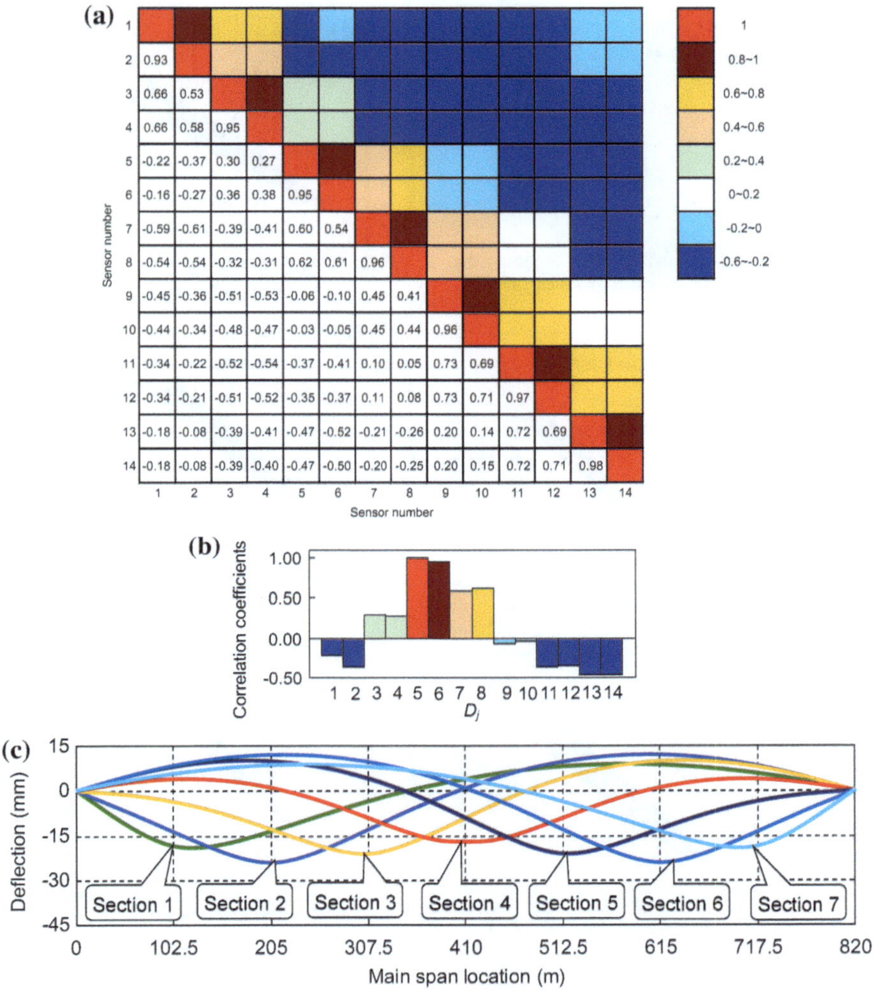

Fig. 8.6 Correlation coefficients $\rho_{D_iD_j}$ of the vertical deflection for the 14 sensors: **a** correlation coefficient matrix, **b** $\rho_{D_5D_j}$ and **c** vertical deflection influence lines

to section 3. The deflections of sections 3, 5, and 7 are −20.8, 4.1, and 8.1 mm, respectively. In this case, deflection of section 7 is larger than that of section 5. The above correlation analysis indicates that the deflection data collected by the CPS is consistent to the mechanical behaviors of the suspension bridge.

Furthermore, the linear regression models of the upstream and downstream sensors are presented in Fig. 8.7. In this figure, the regression lines are approximately crossing the original point. The slopes of the straight lines in the figure can reflect the size relationship between the upstream and downstream deflections in the statistical sense. Taking D_2 versus D_1 as an example, the slope of the regression line is 1.30. The

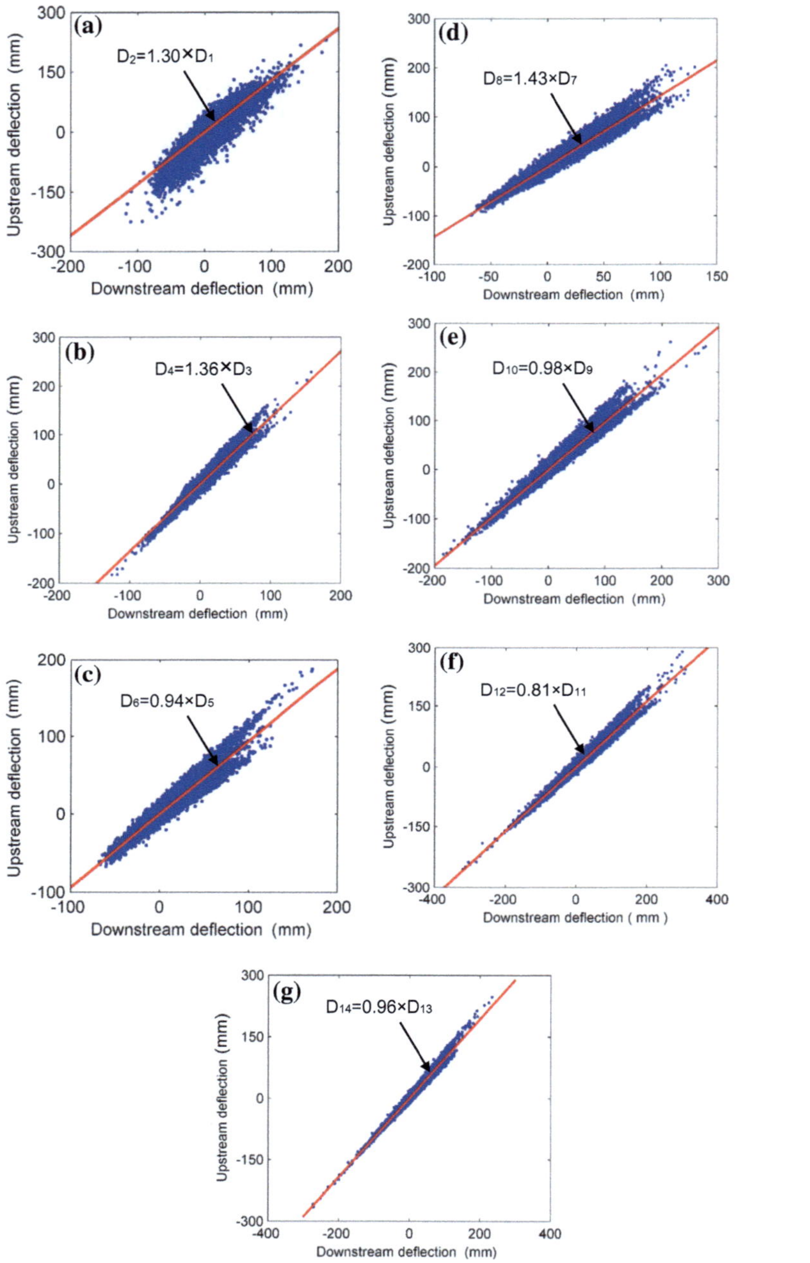

Fig. 8.7 Linear regression models of the measurements from the upstream and downstream sensors: **a** D_2 versus D_1, **b** D_4 versus D_3, **c** D_6 versus D_5, **d** D_8 versus D_7, **e** D_{10} versus D_9, **f** D_{12} versus D_{11} and **g** D_{14} versus D_{13}

regression model indicates that D_2 is statistically 30% more than D_1. Hence, sensor S2 is chosen to represent section 1 to obtain the most unfavorable deflection of this section. Similarly, based on the slopes in Fig. 8.7, a decision can be made that sensors S4, S5, S8, S9, S11, and S13 are chosen to represent sections 2–7, respectively.

8.3.2 GEVD-Based Extreme Value Distribution

In this section, the data of sensor S5, representing the deflection of section 3, is employed as an example to illustrate the strategy for the threshold selection (Fig. 8.8).

Note that the data in Fig. 8.8 is the absolute values of the extracted local extremes and the parameters of the GPD are estimated by using the method of the maximum likelihood estimation (MLE) [36]. The MRLP is shown in Fig. 8.9 and the goodness-of-fit results are shown in Fig. 8.10. H_0 in Fig. 8.10 indicates a failure-to-reject the null hypothesis and H_1 indicates the rejection of the null hypothesis. The significance level of the hypothesis tests is 0.05. In these figures, there are three key points. The first point indicates that the threshold u is 58 mm, above which the exceedances don't reject to yield the GPD. The second point represents that the threshold u is 63 mm, above which it can be approximately observed that the MEF begins to decrease linearly with u. The highest p-value of the goodness-of-fit tests can be seen at the third point, where the threshold u is 71 mm.

Furthermore, the Quantile-Quantile plots of the GPD with the three thresholds are presented in Fig. 8.11. When the thresholds are 58, 63 and 71 mm, there are 1144, 608 and 241 exceeding points in Fig. 8.12a–c, respectively. In the figure, many data points can be observed to deviate obviously from the straight lines when the thresholds are 58 mm or 63 mm. When the threshold is 71 mm, almost all the data points distribute close to the straight line. Figure 8.11 suggests that the threshold of 71 mm can allow the best GPD fitting for the exceeding points. Hence, the proper GPD threshold for the data of sensor S5 is 71 mm according to the proposed threshold selection strategy in step 3.

By using the MLE the shape parameter ξ and scale parameter σ of the data of sensor S5 are 0.116 and 13.2 with the threshold u being 71 mm, respectively. The number of the exceeding points is 241 during the monitoring period of 25 days.

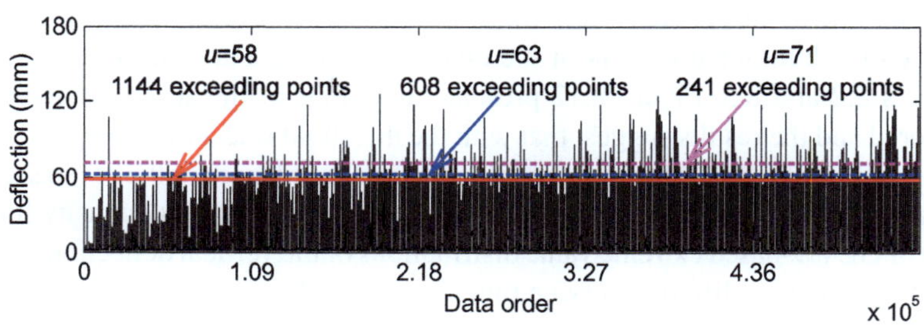

Fig. 8.8 Vertical deflection data of sensor S5 (section 3)

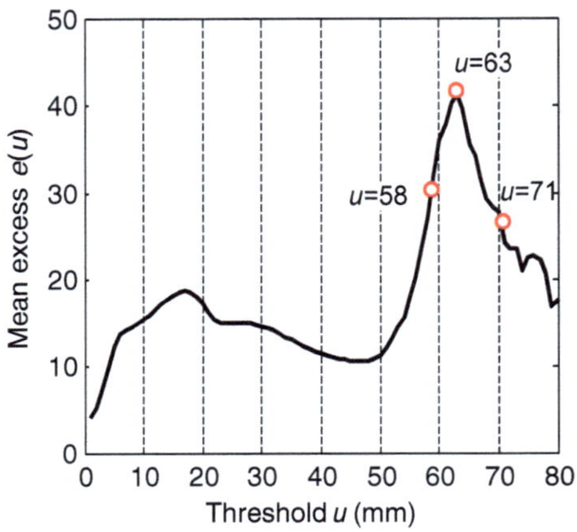

Fig. 8.9 Mean residual life plot of section 3 (sensor S5)

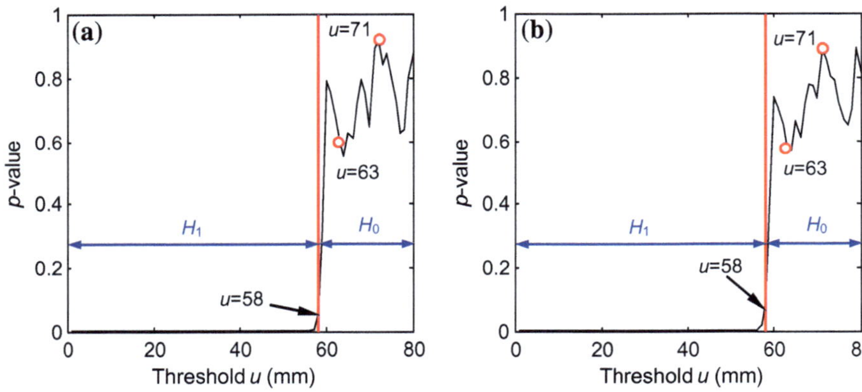

Fig. 8.10 Goodness-of-fit results: **a** Anderson–Darling test and **b** Crámer-von Mises test

Hence, the annual incidence rate of the exceedances λ_u is 3518.6 ($365 \times 241/25$). Then the parameters of the GEVD-based extreme value distribution can be computed according to Eq. (8.17). Supposing the service time T is 100 years, the GEVD parameters $\tilde{u}$, $\tilde{\sigma}$ and $\tilde{\xi}$ are 458.2, 58.2 and 0.116, respectively. Along the same line, the distribution parameters of the other sensors' data are estimated and listed in Tables 8.2 and 8.3. Figure 8.12 presents the GPDs of the seven sections being monitored and the results suggest that with the threshold selection strategy proposed in this study, the GPD is appropriately fitted to the exceedances of the seven sections' vertical deflection. In addition, Fig. 8.12 also shows the probability density function (PDF) of GEVD-based extreme value distributions of the vertical deflections at seven sections throughout different service times, namely 50 years, 100 years (design life of NSB) and 150 years.

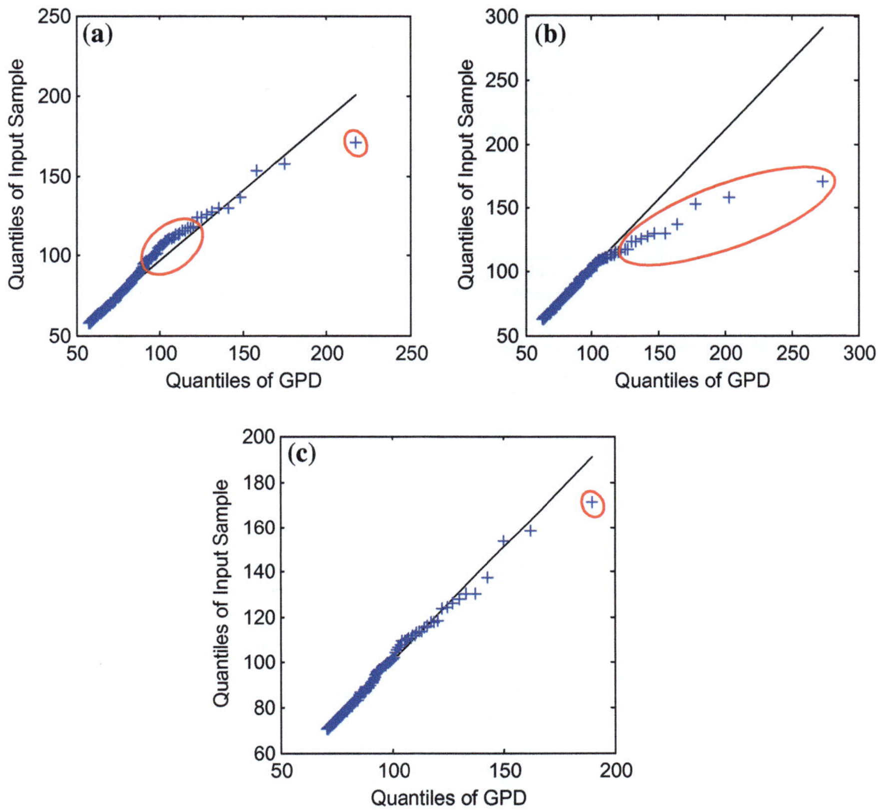

Fig. 8.11 Quantile-quantile plots with various thresholds: **a** threshold $u = 58$ mm, **b** threshold $u = 63$ mm and **c** threshold $u = 71$ mm

8.3.3 Reliability Analysis

The section deflection limit state function was previously presented in Eq. (8.19). In this equation, R_i is the deflection limit of section i. Two levels of the deflection limit are proposed in this study. The first-level deflection limit is the allowable deflection specified by the latest Chinese suspension bridge design specification [2]. Each section has the same first-level deflection limit, the value of which is 3280 mm (1/250 of the main span).

The second-level deflection limit is the calculated deflection under the design vehicle loads. Obviously, the second-level deflection limit is a higher vertical deflection requirement than the first-level one. The highway class-I lane load in Fig. 7.16, specified by the general code for design of highway bridges [37], is the design vehicle load of the NSB. The deflection for each section under the design vehicle loads can be obtained by applying the lane loads to the influence lines shown in Fig. 8.6c. In addition, two reduction factors of the design deflection should be involved. Accord-

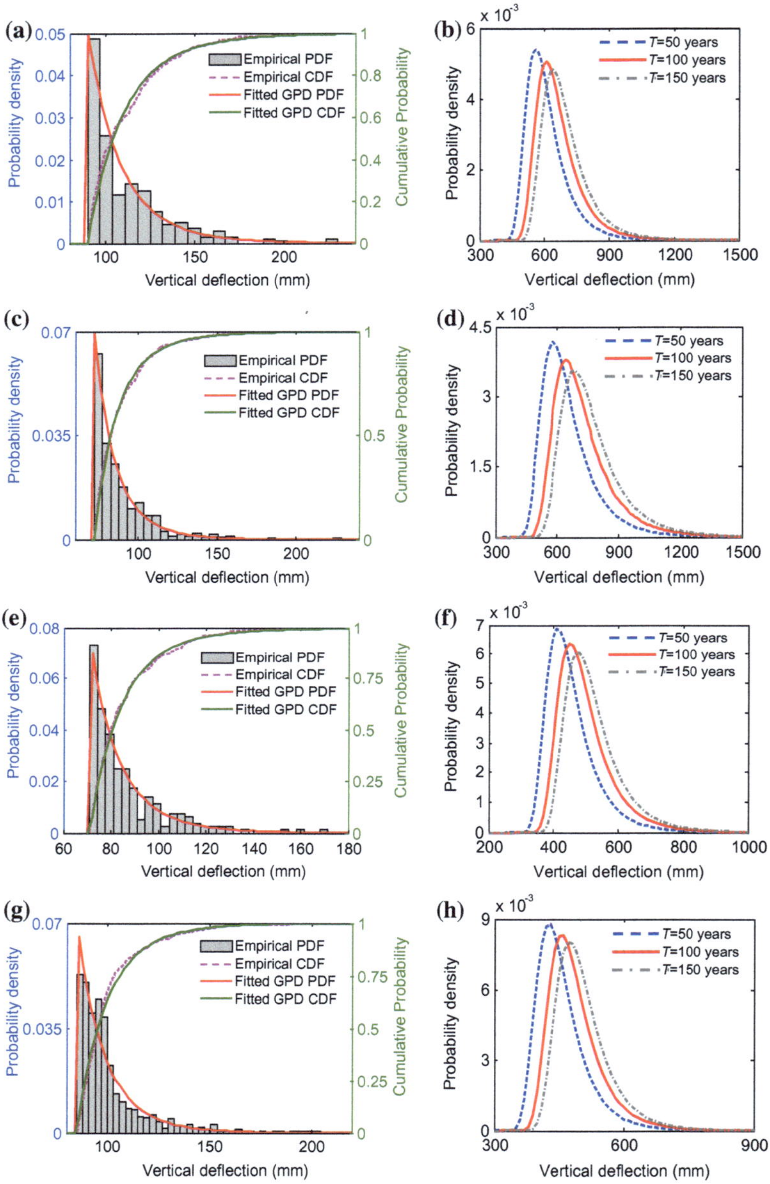

Fig. 8.12 GPD models and extreme values distributions: **a** GPD of section 1, **b** GEVD of section 1, **c** GPD of section 2, **d** GEVD of section 2, **e** GPD of section 3, **f** GEVD of section 3, **g** GPD of section 4, **h** GEVD of section 4, **i** GPD of section 5, **j** GEVD of section 5, **k** GPD of section 6, **l** GEVD of section 6, **m** GPD of section 7 and **n** GEVD of section 7

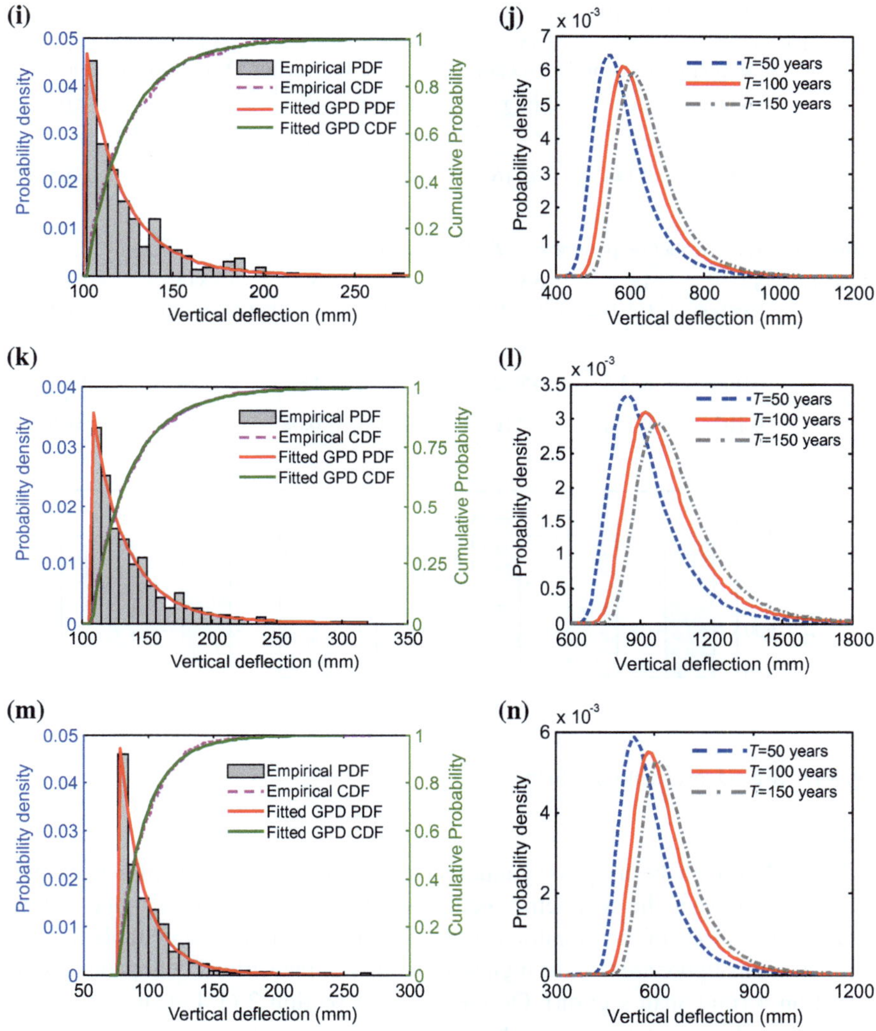

Fig. 8.12 (continued)

ing to the general code [37], the multi-lane reduction factor is 0.55 on account of six lanes in the NSB. The span reduction factor is 0.94 because the main span is within the range from 800 to 1000 m. Figure 8.13 presents the section deflections under the design loads. For example, the second-level deflection limits of section 8.2 and section 4 are 1935 and 1352 mm, respectively.

The PDF of the extreme value distribution of L_i is shown in Fig. 8.12. The vertical deflection reliability of the seven equipped sections can be obtained based on Eq. (8.20). The reliability indices for the seven sections at $T = 100$ years are shown in

Table 8.2 GPD distribution parameters

Parameters	*S*2 (section 1)	*S*4 (section 2)	*S*5 (section 3)	*S*8 (section 4)	*S*9 (section 5)	*S*11 (section 6)	*S*13 (section 7)
σ	19.575	14.3	13.2	14.1	21.5	26.9	19.0
μ	90	72	71	85	102	108	77
ξ	0.102	0.144	0.116	0.080	0.080	0.113	0.095

Table 8.3 GEV distribution parameters (T=100 years)

Parameters	*S*2 (section 1)	*S*4 (section 2)	*S*5 (section 3)	*S*8 (section 4)	*S*9 (section 5)	*S*11 (section 6)	*S*13 (section 7)
$\tilde{\sigma}$	73.4	98.2	58.2	44.2	60.4	120.0	67.4
$\tilde{\mu}$	616.3	655.8	458.2	459.1	589.1	934.0	589.1
$\tilde{\xi}$	0.102	0.144	0.116	0.080	0.080	0.113	0.095

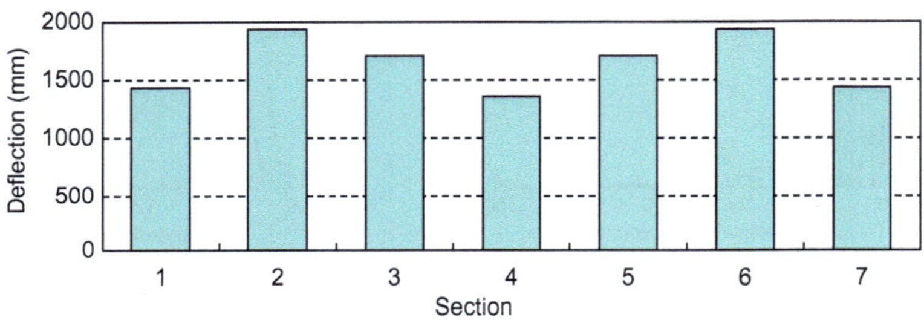

Fig. 8.13 Second-level deflection limits

Fig. 8.14. Although the bridge is symmetric, it is found that there are differences in the reliabilities between the symmetric sections owing to the differences in the extreme values distributions of the monitored vertical deflection as shown in Table 8.3. The findings in this figure are that the highest reliability of the vertical deflection is found at section 4 (mid-span section). On the contrary, section 2 (1/4 span section) and section 6 (3/4 span section) have the lowest reliabilities. The reliabilities with the second-level deflection limits is less than that of the first-level limit. Taking section 4 as an example, the reliability index decreases from 6.3 to 4.4 when its deflection limit decreases from 3280 to 1352 mm.

Figure 8.15 presents the section reliability indices versus service time T from 20 years to 150 years. Based on the section reliability results, the vertical deflection system reliability of the NSB can be derived. According to Eqs. (8.21) and (8.22), the correlation coefficient ρ_{ij} of the failure modes for section i and section j is the premise for the determination of the system reliability. As shown in Eq. (8.23), $\rho_{L_iL_j}$, which is the correlation coefficient of the extreme vertical deflection for section i and section j, can be an alternative to ρ_{ij}. L_i, a random variable yielding to the GEVD, represents the extreme vertical deflection for section i in the limit state function

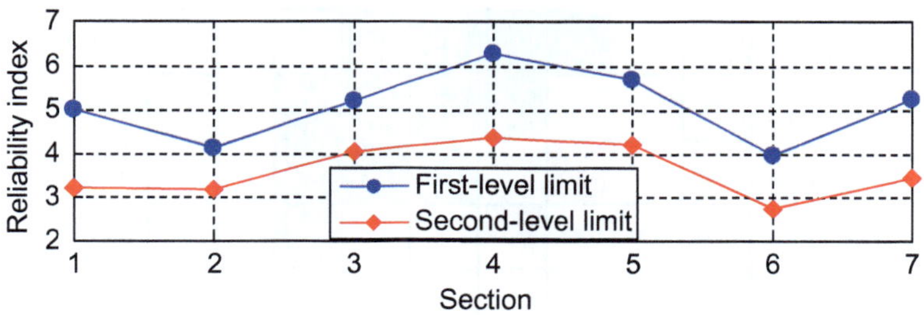

Fig. 8.14 Section reliability indices in 100-year service time

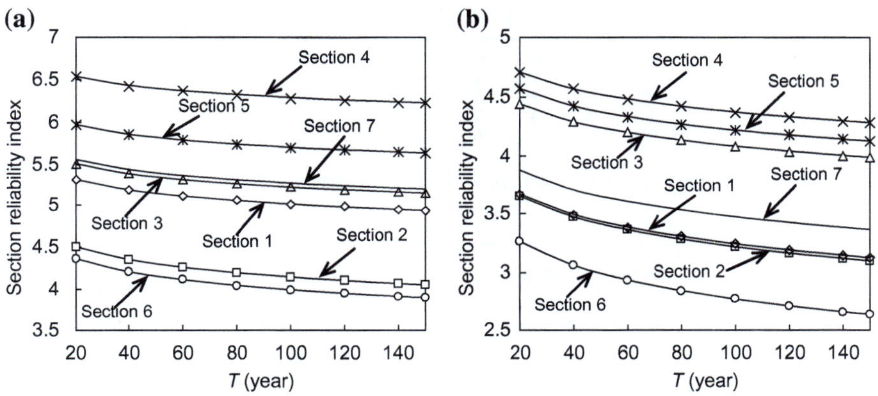

Fig. 8.15 Section reliability indices versus service time T: **a** first-level deflection level and **b** second-level deflection level

[shown in Eq. (8.19)]. As shown earlier, Fig. 8.6 presents the correlation coefficients $\rho_{D_iD_j}$ of the original traffic-induce vertical deflection. Since it is nearly impossible to obtain $\rho_{L_iL_j}$ directly, it is assumed to be approximate to $\rho_{D_iD_j}$. The correlation coefficient matrix $\rho_{L_iL_j}$ is shown in Fig. 8.16. For example, $\rho_{L_1L_2}$ in Fig. 8.16 is equal to $\rho_{D_2D_4}$ because the data of sensors S2 and S4 are used to represent the vertical deflection of section 1 and section 2, respectively. The series system reliability of the seven sections can be derived according to Eq. (8.21). Figure 8.17 presents the reliability bounds of the bridge system. The results in the figure indicate that the upper and lower bounds nearly equal to each other.

With the first-level deflection limit, the upper and lower bounds of system reliability are the same to four decimal places and $\beta_{\text{upper}}=\beta_{\text{lower}}=3.8899$ when the service time T is 100 years (design reference period). When the service time T is 150 years, the reliability decreases to 3.8. With the second-level deflection limits, when the service times are 100 and 150 years, the system reliability indices are 2.6 and 2.5, respectively. According to ISO 13822 [38], the target reliability index is chosen as 1.5. Hence, the system reliability index of the traffic-induced vertical deflection of the NSB is above the target reliability index during the 150-year service time.

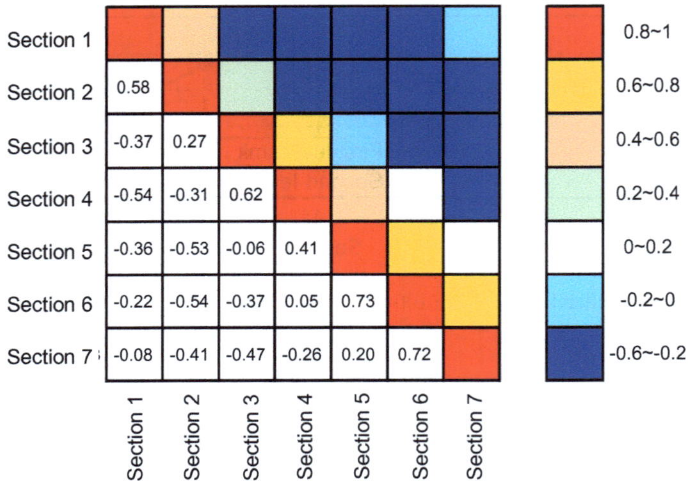

Fig. 8.16 Correlation coefficient matrix $\rho_{L_i L_j}$ of extreme vertical deflection for the seven sections

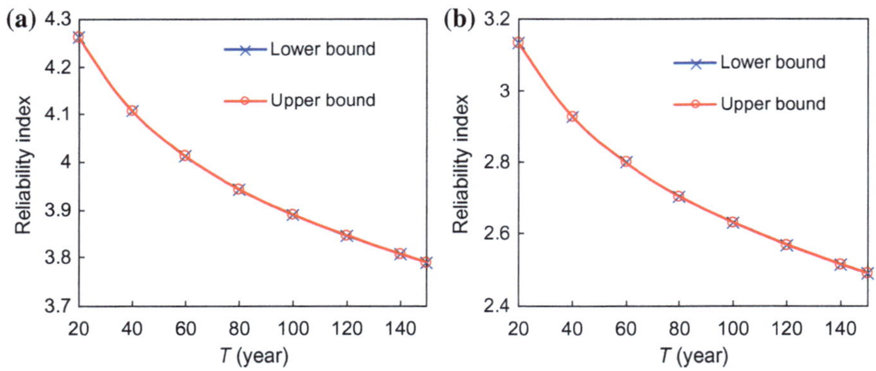

Fig. 8.17 System reliability bounds: **a** first-level deflection level and **b** second-level deflection level

8.4 Summary

This chapter presents a reliability assessment framework of vertical deflection for long-span bridges by joint use of theories of extreme value and system reliability with monitored measurements from SHMS. The main contributions and findings are as follows:

1. The wavelet decomposition can be used effectively for eliminating the temperature effects from the original measurements of the vertical deflection for long-span bridges. The local extremes of traffic-induced vertical deflection can be extracted from the remaining data by using the three-point comparison algo-

rithm. The threshold selection strategy, which is developed based on the mean residual life plot and goodness-of-fit tests, can result in the best fitting results when the GPD is fitted to the extracted local extremes.

2. Based on the theory of filtered Poisson process, when GPD is adopted as the distribution of the POT vertical deflection, the extreme value distribution yields to the GEVD. With the extreme value distributions of the vertical deflections, the deflection limit state function of each instrumented bridge section can be obtained. Then, the series system reliability theory is borrowed to establish a quasi-system reliability model for a long-span bridge with serval in instrumented sections.
3. The relationship between the reliabilities of the instrumented sections of long-span bridges can be approximated as series system. The correlation coefficients between the serviceability limit states of the instrumented sections are assumed to equal the correlation coefficients of the vertical deflection with the contribution from temperature effects being eliminated. Two levels of the deflection limits are proposed for reliability calculation of each bridge section. The first-level limit is the allowable deflection specified by the bridge design code. The second-level deflection limit, which is defined as the calculated deflection under the design vehicle loads, is a higher deflection requirement than the first-level limit.
4. Based on the reliabilities of the instrumented sections, the reliability bounds of the quasi-system composed of multi instrumented sections can be finally derived. For the prototype bridge, the upper and lower bounds of the system reliability are almost the same. When the service time is set to 150 years, the system reliabilities indices are 3.8 and 2.5 for the first-level limit and the second-level limit, respectively. It can be easily found that the system reliabilities are more than the target reliability specified by the ISO 13822 standard. Given the above, the main contribution of this study is a new deflection assessment framework, which is established based on extreme values analysis, system reliability, and field measurements. Note that this framework is not limited to suspension bridges. The application of this framework can be extended to all types of long-span bridges.

References

1. AASHTO. AASHTO LRFD bridge design specifications. 8th ed; 2017.
2. Ministry of Transport of the People's Republic of China. Specifications for design of highway suspension bridge JTG/T D65-05-2015. Beijing: China Communications Press; 2015.
3. Ministry of Transport of the People's Republic of China. Guidelines for design of highway cable-stayed bridge JTG/T D65-01-2007. Beijing: China Communications Press; 2007.
4. BSI. Steel, concrete and composite bridges—Part 2: specification for loads: BS 5400-2:2006: 2006.
5. Feng DM, Feng MQ. Model updating of railway bridge using in situ dynamic displacement measurement under trainloads. J Bridge Eng. 2015;20(12):04015019.
6. Feng DM, Feng MQ. Computer vision for SHM of civil infrastructure: from dynamic response measurement to damage detection—a review. Eng Struct. 2018;156:105–17.

7. Liu Y, Deng Y, Cai CS. Deflection monitoring and assessment for a suspension bridge using a connected pipe system: a case study in China. Struct Control Health Monit. 2015;22(12):1408–25.
8. Feng DM, Feng MQ. Experimental validation of cost-effective vision-based structural health monitoring. Mech Syst Signal Process. 2017;88:199–211.
9. Feng DM, Feng MQ, Ozer E, Fukuda Y. A vision-based sensor for noncontact structural displacement measurement. Sensors. 2015;15(7):16557–75.
10. Ellingwood BR. Reliability-based condition assessment and LRFD for existing structures. Struct Saf. 1996;18(2–3):67–80.
11. Bhattacharya B, Li DG, Chajes M, Hastings J. Reliability-based load and resistance factor rating using in-service data. J Bridge Eng. 2005;10(5):530–43.
12. Bhattacharya B, Li D, Chajes M. Bridge rating using in-service data in the presence of strength deterioration and correlation in load processes. Struct Infrastruct Eng. 2008;4(3):237–49.
13. Messervey TB, Frangopol DM, Casciati S. Application of the statistics of extremes to the reliability assessment and performance prediction of monitored highway bridges. Struct Infrastruct Eng. 2011;7(1–2):87–99.
14. Li SL, Zhu SY, Xu YL, Chen ZW, Li H. Long-term condition assessment of suspenders under traffic loads based on structural monitoring system: application to the Tsing Ma Bridge. Struct Control Health Monit. 2012;19(1):82–101.
15. Gokce HB, Catbas FN, Gul M, Frangopol DM. Structural identification for performance prediction considering uncertainties: case study of a movable bridge. J Struct Eng. 2013;139(10):1703–15.
16. Kaloop MR, Li H. Sensitivity and analysis GPS signals based bridge damage using GPS observations and wavelet transform. Measurement. 2011;44(5):927–37.
17. Wu BJ, Li ZX, Chan THT, Wang Y. Multiscale features and information extraction of online strain for long-span bridges. Smart Struct Syst. 2014;14(4):679–97.
18. Castillo E. Extreme value theory in engineering. New York: Academic Press; 1988.
19. Ditlevsen O. Traffic loads on large bridges modeled as white-noise fields. J Eng Mech ASCE. 1994;120(4):681–94.
20. Coles S. An introduction to statistical modeling of extreme values. Great Britain: Springer; 2001.
21. OBrien EJ, Schmidt F, Hajializadeh D, Zhou XY, Enright B, Caprani CC, Wilson S, Sheils E. A review of probabilistic methods of assessment of load effects in bridges. Struct Saf. 2015; 53:44–56.
22. Bailey SF, Bez R. Site specific probability distribution of extreme traffic action effects. Probab Eng Mech. 1999;14(1–2):19–26.
23. Crespo-Minguillon C, Casas JR. A comprehensive traffic load model for bridge safety checking. Struct Saf. 1997;19(4):339–59.
24. Balkema AA, De Haan L. Residual life time at great age. Ann Probab. 1974;2(5):792–804.
25. Pickands J. Statistical inference using extreme order statistics. Ann Stat. 1975;3(1):119–31.
26. Scarrott C, MacDonald A. A review of extreme value threshold estimation and uncertainty quantification. Revstat Statist J. 2012;10(1):33–60.
27. Langousis A, Mamalakis A, Puliga M, Deidda R. Threshold detection for the generalized Pareto distribution: review of representative methods and application to the NOAA NCDC daily rainfall database. Water Resour Res. 2016;52(4):2659–81.
28. Choulakian V, Stephens MA. Goodness-of-fit tests for the generalized Pareto distribution. Technometrics. 2001;43(4):478–84.
29. Zhou XY, Schmidt F, Toutlemonde F, Jacob B. A mixture peaks over threshold approach for predicting extreme bridge traffic load effects. Probab Eng Mech. 2016;43:121–31.
30. Ministry of Transport of the People's Republic of China. Unified standard for reliability design of highway engineering structures GB/T 50283-1999. Beijing: China Planning Press; 1999.
31. Snyder DL. Random point process. New York: Wiley; 1975.
32. Roth M, Buishand TA, Jongbloed G, Tank AMGK, van Zanten JH. A regional peaks-over-threshold model in a nonstationary climate. Water Resour Res. 2012;48(11):W11533.

33. Wi S, Valdes JB, Steinschneider S, Kim TW. Non-stationary frequency analysis of extreme precipitation in South Korea using peaks-over-threshold and annual maxima. Stoch Env Res Risk Assess. 2016;30(2):583–606.
34. Ditlevsen O. Narrow reliability bounds for structural systems. J Struct Mech. 1979;7(4):453–72.
35. Estes AC. A system reliability approach to the lifetime optimization of inspection and repair of highway bridges. Boulder, CO: University of Colorado at Boulder; 1997.
36. Zhou GD, Yi TH, Chen B, Zhang H. A generalized Pareto distribution-based extreme value model of thermal gradients in a long-span bridge combining parameter updating. Adv Struct Eng. 2017;20(2):202–13.
37. Ministry of Transport of the People's Republic of China. General code for design highway bridges and culverts JTG D60-2004. Beijing: China Communications Press; 2004.
38. International Organization for Standardization. ISO 13822: basis for design of structures—assessment of existing structures. 2003.

Part 5

Chapter 9
Tension Force Monitoring of Main Cable

9.1 Introduction

As a type of critical infrastructure, long-span bridges are usually vital to the local traffic around the world. Structural health monitoring (SHM) techniques have often been applied to assess the conditions of these important structures [1–4]. Due to the complexity of the bridge structures, associated high cost of SHM, and performance limitations of sensors, many challenges remain in terms of how to conduct SHM and interpret the data more effectively and economically [3]. For suspension bridges, main cable is one of the most critical components and some efforts have been carried out to evaluate its performance condition through applying SHM techniques [5, 6]. For instance, a model has been developed to link the horizontal tension of a main cable and measured natural frequencies [6].

Great challenges still exist in monitoring main cables' tension forces during in-service time. It is difficult to directly measure the tension force of a main cable through the means of conventional sensing techniques due to the large cross-section size of the main cable (formed with thousands of parallel high-strength steel wires) and huge tension force values. Other studies adopted the electromagnetic (EM) stress sensor to obtain the tension forces of a main cable based on the correlation of the relative permeability and tensile stress [7–9]. Due to the size limit and relatively high cost of the sensor, EM sensors were primarily used in measuring tension forces of arch bridge hangers, post-tensioned tendons [7, 8] or external tendons of the bridges' box girders [9]. In fact, the studies on long-term in-service tension monitoring of main cables of suspension bridges have been rarely reported.

Instead of directly measuring the tension force of the entire main cable, an alternative is to measure tension forces of a series of strands in the anchor span, which is from the saddle to anchor system in the anchorage. The measured tension forces of strands will be integrated to obtain the total force of the entire main cable. There are several approaches to measure the tension forces of individual strands. Firstly, the tension forces can be derived from the measured basic frequencies of individual

Y. Deng and A. Li, *Structural Health Monitoring for Suspension Bridges*,
https://doi.org/10.1007/978-981-13-3347-7_9

cable strands [10, 11]. Such a method was inspired by the vibration-based tension identification theories for stayed cables [12] and was used to control the tension forces of the strands during the construction period. Along the same line, Zhang et al. [10] established finite element models of anchor span strands of Taizhou Suspension Bridge. Then the mapping relationships between vibration frequencies and the tension forces of the strands in the anchor span were derived by finite element analysis. Wang et al. [11] further improved the formula linking the vibration frequency with tension force of the anchor span strands by considering the influence of the anchor device. Recently, vision sensors have used to identify cables' dynamic characteristics and then the tension forces of the cables can be estimated [13, 14]. This vision-based technique is a new option of vibration-based technique for cables' tension force estimation. However, the vibration-based technique cannot carry out real-time and continuous data acquisition. Thus, the tensions of the strands obtained by this technique are discrete in time. The second technique for strand tension monitoring is Fiber Bragg Grating (FBG) force testing ring invented by Nan [15]. This kind of device can provide precise measurements but with relatively high cost. In addition, there are some emerging techniques including FBG-based smart strand [16, 17] and guided ultrasonic or stress waves [18–22]. As of now, these techniques, however, have been investigated in laboratories and rarely tested in the field monitoring practices of bridges.

Therefore, to develop a highly reliable, cost-effective, and precise system for tension monitoring of main cable is desirable. Conventional vibrating strings transducer (VST), a kind of reliable, low-cost, and technology-mature sensor for pressure measuring, is used in monitoring of main cable. VST has already been well developed [23, 24] and applied in civil engineering, such as tension monitoring of main cable strands during construction [25], strain monitoring of building columns during construction [26] and temperature and strain monitoring of concrete bridges during construction [27]. Due to its reliability and low-cost, VST has been used in construction monitoring of structures with good performance. The feasibility of adopting VST on measuring tension forces of cable strands can be partially verified by a study by Huang and Mu [25]. In their study, VSTs were used to measure and control the strands' tension forces in the anchor span during the construction of a long-span suspension bridge. However, there is no reported study about applying the VSTs on the long-term monitoring of suspension bridges during service time.

A long-term monitoring strategy of applying VSTs on a long-span suspension bridge to collect in-service main cable force data is developed. After the VST transducers measure the forces of a series of separate cable strands, the tension of the whole main cable is further derived approximately. With the monitored data, the performance assessment of main cable forces based on the measured data is also demonstrated on a prototype bridge. One-day and seven-month monitoring data are processed and analyzed to provide some insights of main cable tension forces. The uniformity of strands' tension forces and the main cable's safety factor related to load-bearing capacity are also evaluated.

9.2 Tension Force Monitoring System of the NSB

The prototype bridge in this chapter is the Nanxi Suspension Bridge (NSB) [28]. The strands in NSB's main cable are prefabricated parallel wire strands (PPWS). There are 87 strands in each main cable passing through two cable pylons from the Yibin anchorages to the Luzhou anchorages and two more strands are added to each main cable in side spans. Each strand is composed of 127 high-strength galvanized steel wires, of which the diameter and nominal tensile strength are 5.1 mm and 1670 MPa, respectively. Thus, the cross-section area of each main cable in the anchor span is 230,783 mm^2 (89×2593 mm^2). Figure 9.1a shows the saddle of the Yibin downstream anchorage. Each separate strand is anchored individually by using a single anchoring system as shown in Fig. 9.1b.

In main cable's anchor span, the separation of cable strands provides an opportunity to evaluate the main cable's tension by monitoring several cable strands' tension forces. As shown in Fig. 9.1c, d, two VSTs were mounted between the socket and nuts of pull rods for a single cable strand. The transducers can measure the compression forces between the nuts and socket. Obviously, the strand's tension force equals the sum of two transducers' compression forces.

When a pressure is applied on a VST sensor, the deformation of the sensor causes variations of the tension forces of the strings inside the sensor, which lead to changes of the strings' frequencies [23, 24]. By measuring the frequencies of the stretched strings inside a VST sensor, the pressure can be computed through the pressure–frequency relationship. The key parts of vibrating string transducer consist of a thin circular plate (radius a, thickness h) and string, see Fig. 9.2. One end of the string of length L_0 is connected to the center of the plate o. The other end of the string is moved along its line in a displacement τ_0 before being fixed to point o_1. The in-plane uniform tension p_0 is applied around the edge of the plate. The string, connected to the center of the plate, is pre-stretched. The transverse static pressure q applied on the top surface of the plate brings additional transverse displacement of the plate. The change of the central transverse displacement of the plate w_0 results in variation of both the string's tensile force and natural frequencies. The measured fundamental resonant frequency of the string will, therefore, indicate the static pressure applied on the plate [24].

In the construction period, VSTs were employed to monitor the tension forces during the installation process of cable strands to make sure that the sum of two pull rods' tension forces reaches the design value. After installing cable strands, some transducers have been kept continuing monitoring the cable tension during the service time. Installing VSTs to the separate strands is easy to implement. First, the calculated tension forces can help to determine the measuring range of the transducers. During the construction period, the transducers should be placed to the pull rods shown in Fig. 9.1d. Once the tension forces of the strands reach the design values, the nuts should be tightened immediately and then the transducers can't be removed. Figure 9.3 shows six cable strands, on each of which are instrumented with two transducers.

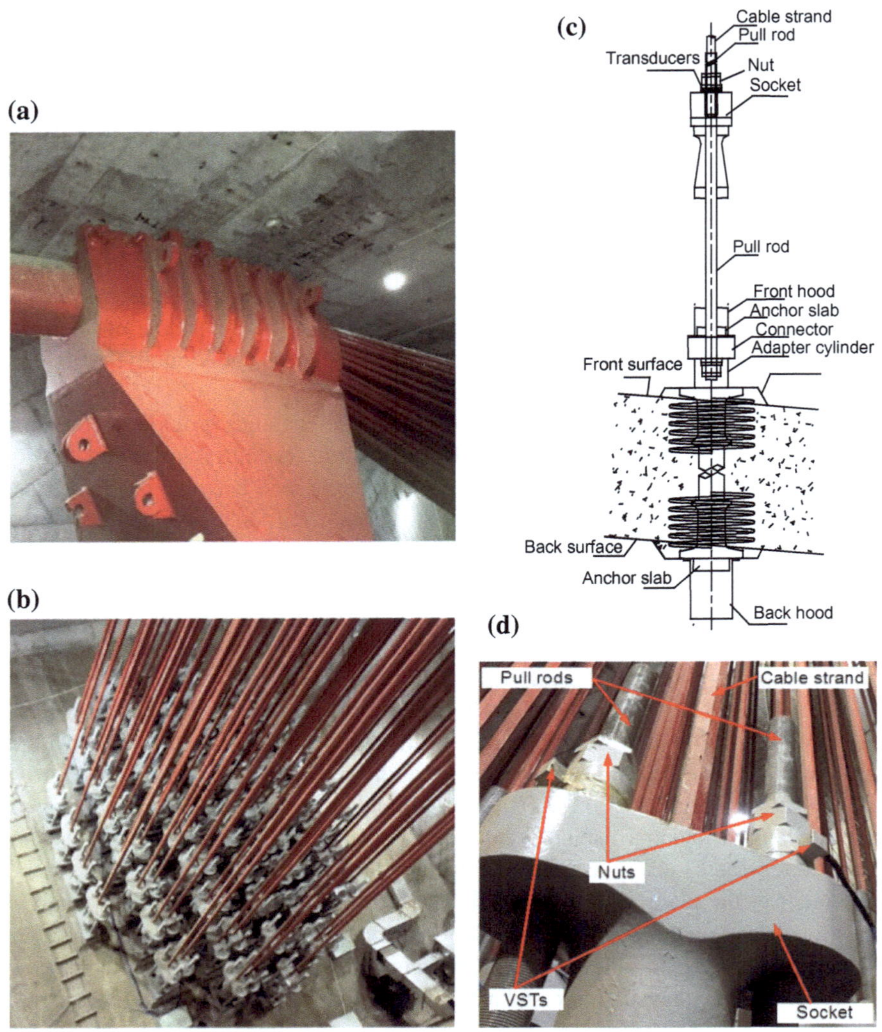

Fig. 9.1 Anchoring system in the anchor span: **a** saddle in Yibin downstream anchorage, **b** front anchor surface in Yibin downstream anchorage, **c** transducers' locations and anchor system and **d** two transducers to monitor one strand's tension force

9.3 Estimation of Main Cable's Tension Force

As shown in Fig. 9.1c, d, transducers are mounted between the socket and pull rods' nuts and each transducer measures the change of the compression force between cable strand's socket and pull rod's nut. The SHMS of the NSB adopted four-string transducers to monitor the tension force of the main cable. Figure 9.2 just shows a string to demonstrate the working mechanism of the VSTs. In practical engineering,

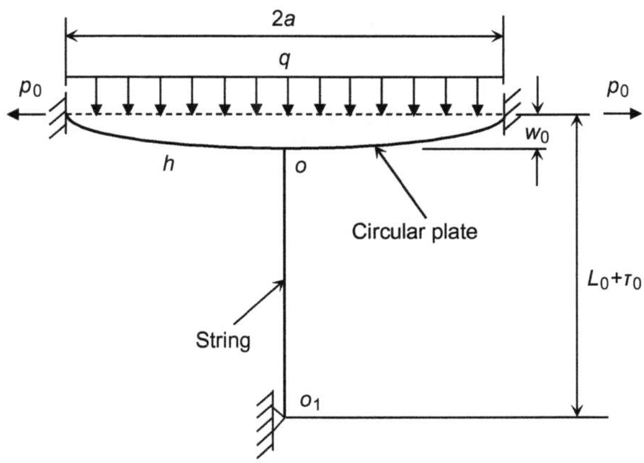

Fig. 9.2 Schematic of the vibrating string transducer

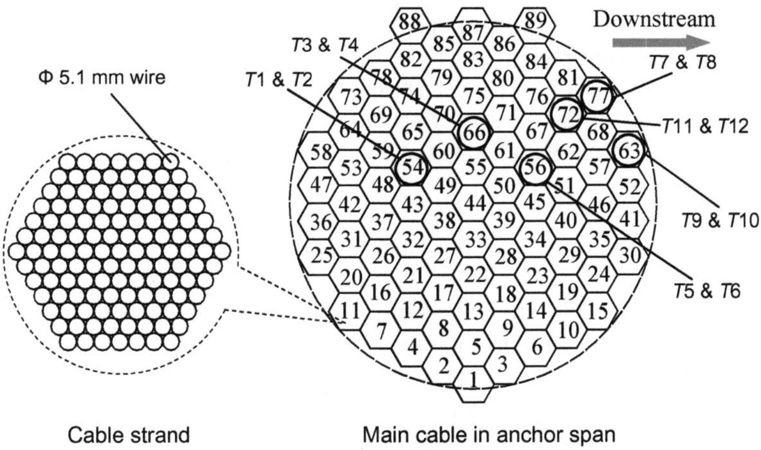

Fig. 9.3 Strand arrangement and transducer placement in main cable of anchor span

usually a multi-string setup of the VSTs is designed to address the possible uneven loading. The transducers' measuring range and sensitivity are 1000 kN and ±1 kN, respectively. According to the basic theory of vibrating string transducer for pressure monitoring [24], the tension force of the main cable can be estimated following the procedures summarized below.

For a single transducer, the current strain of the *j*th string can be calculated based on its fundamental resonant frequency

$$\varepsilon_j = K \times f_j^2, \quad j = 1, 2, 3 \text{ or } 4 \tag{9.1}$$

where ε_j, of which the unit is με, is the strain of the *j*th string. The coefficient K of the strings is 0.00084107 for the type of transducer. f_j is the fundamental resonant frequency of the *j*th string (Hz). The current average strain of 4 strings in a VST can be obtained by

$$\varepsilon = \frac{\varepsilon_1 + \varepsilon_2 + \varepsilon_3 + \varepsilon_4}{4} \tag{9.2}$$

The change of strain $\Delta\varepsilon$ is

$$\Delta\varepsilon = \varepsilon - \varepsilon_0 \tag{9.3}$$

where ε_0 is the initial strain. The strain change after eliminating temperature effect is defined as

$$\Delta\varepsilon_t = \Delta\varepsilon + (t - t_0) \times (C_t - C_x) \tag{9.4}$$

where $\Delta\varepsilon_t$ is the strain change after eliminating temperature effect (με); t is the current temperature (°C); t_0 is the baseline temperature (°C); Temperature–strain coefficient of transducer's body C_t is 11.7 με/°C; Temperature–strain coefficient of transducer's string C_x is 12.2 με/°C. The coefficients K, C_t, and C_x, which are physical characteristics of the transducer's materials, are provided by the manufacture. The compression force N applied to the transducer has a linear relationship with the strain change $\Delta\varepsilon_t$,

$$N = a + b \times \Delta\varepsilon_t \tag{9.5}$$

where a (kN) and b (kN/με) are the correction and calibration coefficients, respectively. Generally, each transducer has its own correction and calibration coefficients. Table 9.1 lists coefficients of all the transducers also provided by the manufacture.

The transducers record one data value per every 2 min. The tension force of main cable T can be approximately estimated as

$$\begin{aligned} T &= 89 \times \frac{\sum T_{Si}}{6} \\ &= 89 \times \frac{\sum_{l=1}^{12} N_l}{6} (i = 54, 56, 63, 66, 77, \text{ and } 72; l = 1, 2, \ldots, 12) \end{aligned} \tag{9.6}$$

$$\begin{aligned} T_{S54} &= N_1 + N_2, T_{S66} = N_3 + N_4, T_{S56} \\ &= N_5 + N_6, T_{S77} = N_7 + N_8, T_{S63} \\ &= N_9 + N_{10}, T_{S72} = N_{11} + N_{12} \end{aligned} \tag{9.7}$$

where T_{Si} represents the tension force of the *i*th cable strand (i=54, 56, 63, 66, 77, and 72). N_l (l=1, 2, 3, …, 12) obtained by using Eq. 9.5 represents the compression force measured by transducer no. l.

Table 9.1 Correction and calibration coefficients

Strand no.	54		66		56		77		63		72	
VST no.	1	2	3	4	5	6	7	8	9	10	11	12
a	−12.7	−42	−9.1	−15	−7.3	−9.7	−6	−9.8	−6.13	−12.3	−8.6	−12
b	1.595	1.591	1.569	1.538	1.584	1.540	1.548	1.536	1.552	1.553	1.552	1.587

9.4 Assessment Methods of Main Cable

9.4.1 Uniformity of Cable Strands' Tension Forces

During the service period, great attention should be paid to the uniformity of cable stands' tension forces in anchor span. If there is significant non-uniformity among the cable strands' tension forces, it means some strands experience more tension forces than what is expected to be. Thus, the performance of the main cable and even whole bridge will be adversely affected. Coefficient of variation (COV) is employed to assess the non-uniformity among the measured tension forces of multiple strands. At a given time, the COV could be calculated as

$$E_T = \frac{\sum T_{Si}}{6} \ (i = 54, 56, 63, 66, 77, \text{and } 72) \tag{9.8}$$

$$\sigma_T = \sqrt{\frac{\sum (T_{Si} - E_T)^2}{6}} \tag{9.9}$$

$$\text{COV} = \frac{\sigma_T}{E_T} \tag{9.10}$$

where E_T and σ_T are the mean value and standard deviation of the current tension forces of the selected strands; T_{si}, which has been defined in Eq. 9.7, is the tension force of strand i. It is obvious that the more COV indicate higher non-uniformity.

9.4.2 Thresholds of Main Cable's Safety Factor

It is well known that the actual loads on the bridge during service period are totally different from the design loads. The suspension bridge may encounter some extreme loads such as overloading trucks and strong winds. Besides, corrosion fatigue due to the adverse operation circumstances may result in cross-section loss of main cable. These unfavorable factors will be harmful to main cable's safety. Hence, the safety factor is often used to evaluate the current load carrying capacity of main cable. Base on Chinese standard JTG TD65-05-2015 [29], the safety factor of a main cable can't be less than 2.5. This provision was determined from the previous studies and engineering experience [30]. The safety factor K_s of a main cable is defined as

$$K_S = \frac{\sigma_n}{\sigma} = \frac{\sigma_n \times A}{T} \geq 2.5 \tag{9.11}$$

where σ_n, of which the value is 1670 MPa, is the nominal tensile strength of main cables' steel wires. σ is the calculated stress of steel wires. The cross-sectional area of the main cable A is 230,783 mm^2. T is the estimated tension force of main cable.

According to JTG D60-2004 [31], the design stress of a main cable is related to dead load, vehicle load, pedestrian load, wind load and temperature effects. According to the bridge design documentation, the calculated tension force of main cable in Yibin anchor span is 134,157 kN. Thus, the calculated stress of steel wires σ is 581 MPa. By using Eq. 9.11 the design safety factor of the main cable is 2.87. Hence, two thresholds of the safety factor are established: the first level of safety factor is 2.87 and the second is 2.5. When the actual safety factor is lower than the first level, it suggests that the operational loads are more than the design loads. When lower than the second level, it suggests that the design code requirement is not met, and serious attention should be paid to the possible performance degradation and safety of main cable.

9.5 Long-Term Data Analysis

9.5.1 One-Day Monitoring

The data measured on December 1, 2015 is selected and processed. Strand $S56$ is taken as the study object to demonstrate the estimation process of main cable's tension force from the original monitoring data. A temperature sensor is also integrated into each transducer to measure the temperature data. The transducers are specifically designed to measure and output the real-time data including strings' frequencies, temperature, and compression force. Table 9.2 lists strings' frequencies and baseline temperatures of Transducer $T5$ (attached to $S56$), which were obtained through product calibration by the manufacturer before delivery. According to Eqs. 9.1 and 9.2, the beginning strain of $T5$ can be obtained.

Figure 9.4a shows the actual temperature measurements by $T5$ in the bridge service stage. Figure 9.4b gives the environmental temperature measured by a sensor installed at the middle of the NSB's deck. There is apparent difference between the environmental temperature and the temperature measured by $T5$. The temperature in Fig. 9.4b shows a typical day-and-night change of air temperature. However, there is little daily variation of the temperature in Fig. 9.4a and it seems that there is no influence of the environmental temperature on the measured temperature by $T5$. The reason is that the ground anchorage, isolated from the outside environment, is in a relatively closed space.

Figure 9.5 gives the measured frequencies of strings in $T5$. It is found that the measured frequencies in Fig. 9.5 are lower than those listed in Table 9.2. The lower

Table 9.2 Initial frequencies and baseline temperatures

f_1	f_2	f_3	f_4	Baseline temperature T_0	Initial strain
1989 Hz	1988 Hz	1990 Hz	1991 Hz	33 °C	3329.0 $\mu\varepsilon$

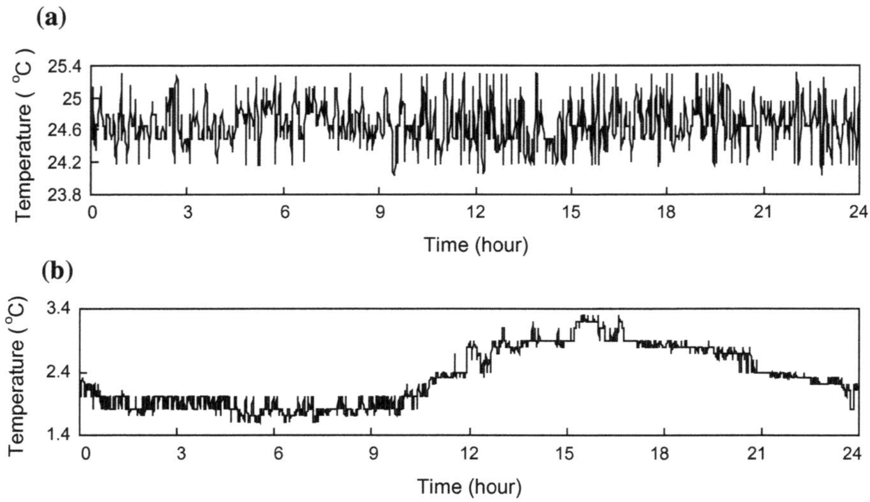

Fig. 9.4 Temperature measurements: **a** temperature measured by transducer *T*5 attached to cable strand *S*56 and **b** environmental temperature

frequencies suggest that there is additional pressure imposed on the transducer [23, 24]. The strings in transducers were pre-stretched and the pressure causes a reduction in string tension. Figure 9.6 shows the transducers' compression forces calculated by using Eqs. 9.4 and 9.5. The compression forces of *T*5 and *T*6 vary in small ranges: from 731 kN to 738 kN and 631 kN to 636 kN, respectively.

The tension force of *S*56 obtained based on compression forces of *T*5 and *T*6, is shown in Fig. 9.7. The results of the other strands are also shown in the figure. Table 9.3 lists statistical characteristics of the selected strands' tension forces measured on Dec. 1, 2015. The data reveals that the largest tension forces were observed in strand *S*66 and *S*77 while Strand *S*54 experienced the least tension force. For each single strand, the daily variation of tension force is less than 15 kN. It can be seen in Table 9.3 that the maximum and minimum mean values are 1393 kN for *S*77 and 1329 kN for *S*54, respectively. The differences among the strands' tension forces are very small.

For every 2 min, Eq. 9.10 is used to calculate a COV of the six cable strands' tension forces. The COV curve of the tension force is plotted in Fig. 9.8a. The COV of Dec. 1, 2015 changed in the range of 1.82–1.84%. The results indicate that the tension force uniformity of the instrumented cable strand is well maintained according to the analysis of 1-day short-term monitoring data.

The main cable's tension forces of Dec. 1, 2015 are estimated by using Eq. 9.6 per every two minutes. As shown in Fig. 9.8b the maximum and minimum tension forces are 122,420 and 121,680 kN, respectively. The safety factors are calculated by using Eq. 9.11. The safety factors of Dec. 1, 2015 in Fig. 9.8c vary in the range between 3.15 and 3.17. Since the safety factors are higher than the first level of safety factor criterion, the monitored tension force of the main cable is within the designed value, exhibiting acceptable performance.

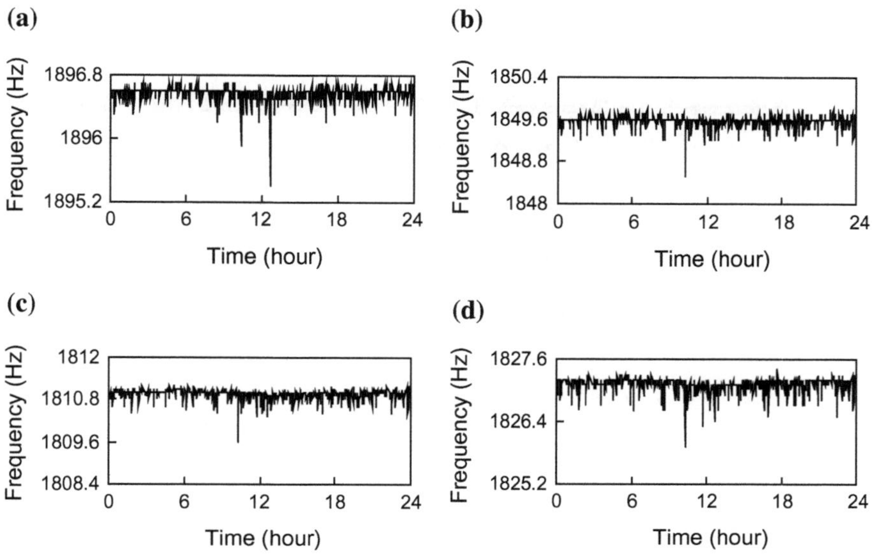

Fig. 9.5 Fundamental resonant frequencies of strings in transducer $T5$: **a** string 1, **b** string 2, **c** string 3 and **d** string 4

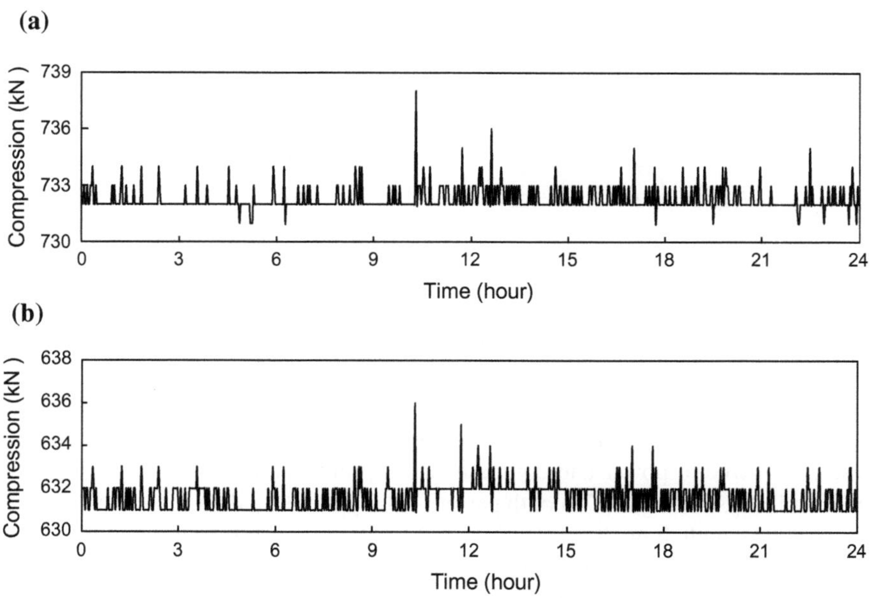

Fig. 9.6 Compression force: **a** $T5$ and **b** $T6$

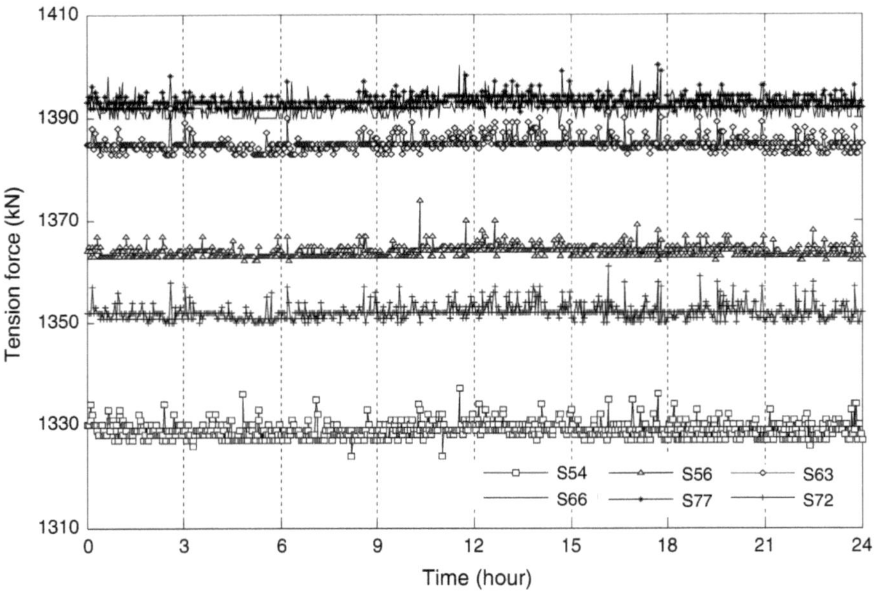

Fig. 9.7 Tension forces of cable strands on Dec. 1, 2015

Table 9.3 Statistical characteristics of tension forces on December 1, 2015

Cable strand	*S*54	*S*66	*S*56	*S*77	*S*63	*S*72
Min value (kN)	1324	1388	1362	1391	1383	1349
Max value (kN)	1337	1400	1374	1400	1393	1361
Mean value (kN)	1329	1392	1364	1393	1385	1352
Change range (kN) (Max − Min)	13	12	12	9	10	12

9.5.2 Long-Term Monitoring

After the NSB was open to traffic, the VST-based system began to monitor the anchor strands' tension forces. A long-term monitoring data from June to December of 2015 is selected. Due to power and communication problems, there was some data loss on several days during the 7-month period. Totally there are data for 162 days which are complete, including 29, 23, 18, 22, 21, 22 and 27 days in June through December of 2015, respectively. For each day, the data of every cable strand's tension force are averaged to obtain a mean value and there are 162 data points in each curve shown in Fig. 9.9.

It can be observed that the tension forces throughout the 7-month period vary randomly and the statistical characteristics of the data in Fig. 9.9 are listed in Table 9.4. The largest tension forces are observed on *S*66 and *S*77 and the least tension force appears on *S*54. The variation of *S*54's tension force, of which the value is 127.9 kN,

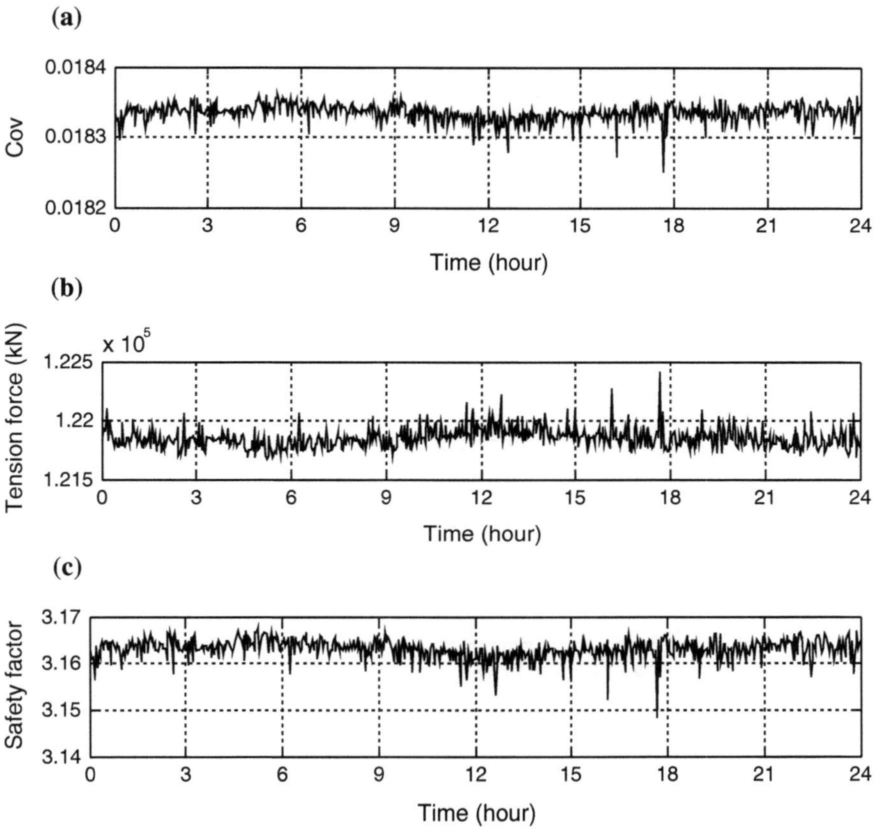

Fig. 9.8 Assessment results on December 1, 2015: **a** COV of cable strands' tension forces, **b** tension forces of main cable in anchor span and **c** safety factor of main cable

is the largest. On the other hand, the variations of the tension forces for the rest of the strands are about 60 kN. The daily COVs throughout the 7 months, varying between 1.25% and 1.35%, are shown in Fig. 9.10a. With small COVs, the uniformity of the monitored tension forces of the instrumented cable strands is deemed excellent based on the long-term monitoring data.

Figure 9.10b gives the daily averaged tension forces of the main cable, which are calculated following Eq. 9.6 based on the tension forces shown in Fig. 9.9. The mean, maximum and minimum tension forces are 121,750, 123,380, and 118,880 kN, respectively. Hence, the variation of the main cable's tension force, of which the value is 4500 kN, is only 3.7% of the mean tension force during the seven months. Such phenomenon is consistent with the existing knowledge about suspension bridges that most of the main cable force comes from permanent loads (e.g. bridge dead loads) and only a small portion is caused by the action from other loads [30, 32].

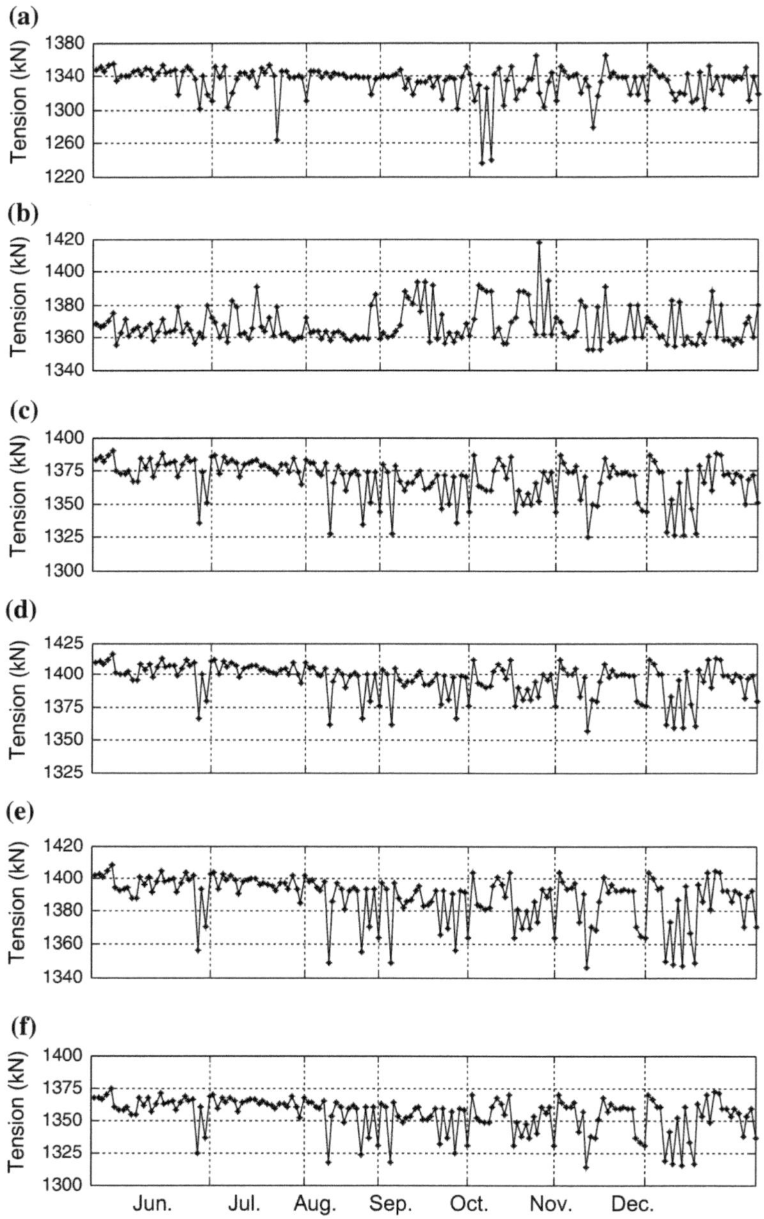

Fig. 9.9 Daily averaged tension forces: **a** *S*54, **b** *S*66, **c** *S*56, **d** *S*77, **e** *S*63 and **f** *S*72

Finally, the safety factor curve during the 7 months is plotted in Fig. 9.10c. The maximum and minimum safety factors are 3.24 and 3.12, respectively. Obviously, the minimum safety factor is not only larger than 2.5—the second level of safety factor

Table 9.4 Statistical characteristics of tension forces in 7 months of 2015

Cable strand	*S*54	*S*66	*S*56	*S*77	*S*63	*S*72
Min value (kN)	1236.7	1352.7	1324.8	1357.5	1346.4	1314.0
Max value (kN)	1364.6	1417.9	1390.4	1416.3	1407.8	1374.4
Mean value (kN)	1333.2	1367.5	1368.6	1396.2	1388.2	1354.5
Change range (kN) (Max − Min)	127.9	65.3	65.7	58.7	61.4	60.3

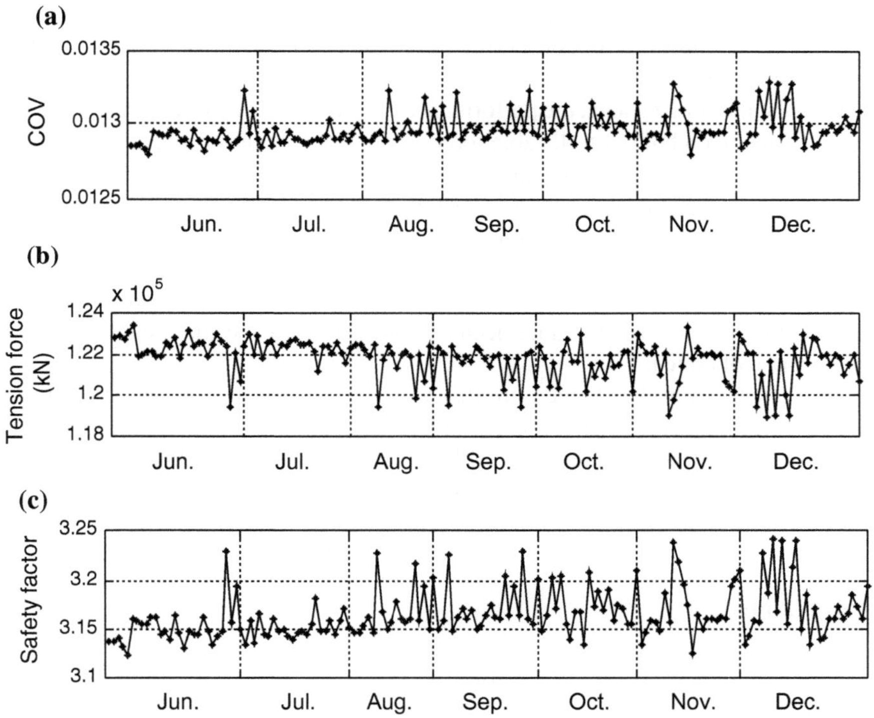

Fig. 9.10 Long-term assessment results: **a** daily COV of instrumented cable strands' tension forces; **b** daily averaged tension forces of main cable in anchor span; **c** daily safety factor of main cable

from the design specification, but also larger than 2.87—the first level of safety factor related to the design actions. The results indicate that the original design, which was carried out based on the design loads or actions, ensures enough capacity of the NSB's main cables during the bridge service period. Because the NSB could still be treated as a new-built suspension bridge, there is no abnormal data in the long-term monitoring. Further long-term monitoring over years can be expected to disclose more information after deterioration occurs.

9.6 Summary

A health monitoring and performance assessment strategy for the main cable force of a suspension bridge was proposed. The proposed technology monitors a series of strands' tensions in the anchor span rather than the entire main cable, which is acceptable for in-service monitoring of main cable's tension. The conventional VSTs were adopted to monitor the main cable's tension force for suspension bridges. Such transducers have many advantages, such as low cost, reliable technology and good accuracy. The low-cost monitoring system based on vibrating string-sensing technology has shown its reliable performance and potential in long-term monitoring of the main cable's tension.

With the monitoring data, the performance evaluation of main cable can be carried out by focusing on both overall cable force and uniformity of forces among different strings. The thresholds of main cable's safety factor were established.

The proposed technology was demonstrated on a prototype long-span suspension bridge and the monitoring data of one-day and seven-month data were analyzed. The results show that the proposed methods can provide approximate estimation of main cables' tension forces. For the prototype suspension bridge, when more transducers are mounted to the strands, more accurate estimation of the main cable's tension force can be obtained.

The lessons from this chapter are that when more data is collected, more information on the in-service performance of the main cable can be revealed. And the conventional sensing technique can still play an effective role in health monitoring for main cables of suspension bridges.

References

1. Chang PC, Flatau A, Liu SC. Review paper: health monitoring of civil infrastructure [J]. Struct Health Monit Int J. 2003;2(3):257–67.
2. Ou J, Li H. Structural health monitoring in mainland China: review and future trends. Struct Health Monit Int J. 2010;9(3):219–31.
3. Webb GT, Vardanega PJ, Middleton CR. Categories of SHM deployments: technologies and capabilities. J Bridge Eng. 2015;20(11):04014118.
4. Deng Y, Liu Y, Feng DM, Li AQ. Investigation of fatigue performance of welded details in long-span steel bridges using long-term monitoring strain data. Struct Control Health Monit. 2015;22(11):1343–58.
5. Deeble Sloane M, Betti R, Marconi G, Hong A, Khazem D. Experimental analysis of a non-destructive corrosion monitoring system for main cables of suspension bridges. J Bridge Eng. 2013;18(7):653–62.
6. Wang J, Liu W, Wang L, Han X. Estimation of main cable tension force of suspension bridges based on ambient vibration frequency measurements. Struct Eng Mech. 2015;56(6):939–57.
7. Ansari F. Sensing issues in civil structural health monitoring. Dordrecht: Springer; 2003.
8. Zhao Y, Wang ML. Fast EM stress sensors for large steel cables. In: Proceedings SPIE 6934, Nondestructive characterization for composite materials, aerospace engineering, civil infrastructure, and homeland security, 09 Mar 2008. Bellingham: SPIE-International Society for Optical Engineering; 2008.

9. Liu L, Chen W, Zhang P, Hu S. Theoretical model and optimization of magnetic stress sensor for measuring steel cable stress. In: Proceedings of SPIE 7544, Sixth international symposium on precision engineering measurements and instrumentation, 28 Dec 2010. Bellingham: SPIE-International Society for Optical Engineering; 2010.
10. Zhang X, Shen R, Tang M, Ye H. Accurate calculation and adjustment methods for cable forces of anchor-span strands for suspension bridges. J Southwest Jiaotong Univ. 2012;47(4):551–7.
11. Wang D, Zhang W, Liu Y, Liu Y. Strand tension control in anchor span for suspension bridge using dynamic balance theory. Lat Am J Solids Struct. 2016;13(10):1838–50.
12. Yu Y, Zhang C, Zhu X, Kang WH, Mao X, Uy B. Vibration based wireless measurement system for bridge cable tension monitoring. Adv Struct Eng. 2014;17(11):1657–68.
13. Feng D, Feng MQ. Experimental validation of cost-effective vision-based structural health monitoring. Mech Syst Signal Process. 2017;88:199–211.
14. Feng D, Scarangello T, Feng MQ, Ye Q. Cable tension force estimates using cost-effective noncontact vision sensor. Measurement. 2017;99:44–52.
15. Nan Q. Application of FBG force testing ring in monitoring cable force of suspension bridge. J Highw Transp Res Dev. 2010;27(3):64–8.
16. Li D, Zhou Z, Ou J. Development and sensing properties study of FRP-FBG smart stay cable for bridge health monitoring applications. Measurement. 2011;44(4):722–9.
17. Kim YS, Sung HJ, Kim HW, Kim JM. Monitoring of tension force and load transfer of ground anchor by using optical FBG sensors embedded tendon. Smart Struct Syst. 2011;7(4):303–17.
18. Lanza di Scalea F, Rizzo P, Seible F. Stress measurement and defect detection in steel strands by guided stress waves. J Mater Civ Eng. 2003; 15(3):219–227.
19. Nucera C, Lanza di Scalea F. Monitoring load levels in multi-wire strands by nonlinear ultrasonic waves. Struct Health Monit Int J. 2011; 10(6):617–629.
20. Deng Y, Liu Y, Feng DM. Monitoring damage evolution of steel strand using acoustic emission technique and rate process theory. J Central South Univ. 2014;21(9):3692–7.
21. Blachowski B, Swiercz A, Gutkiewicz P, Szelazek J, Gutkowski W. Structural damage detectability using modal and ultrasonic approaches. Measurement. 2016;85:210–21.
22. Pnevmatikos NG, Blachowski B, Hatzigeorgiou GD, Swiercz A. Wavelet analysis based damage localization in steel frames with bolted connections. Smart Struct Syst. 2016;18(6):1189–202.
23. Han W, Cleghorn WL. Vibration analysis of pre-stressed pressure sensors using finite element method. Finite Elem Anal Des. 1998;30(3):205–17.
24. Zhou YH. An analysis of pressure-frequency characteristic of vibrating string-type pressure sensors. Int J Solids Struct. 2001;38(40–41):7101–11.
25. Huang PM, Mu YK. Tension control system of suspension bridge anchor cable. J Chang'an Univ (Nat Sci Ed). 2007;27(4):42–5.
26. Choi SW, Kwon EM, Kim Y, Hong K, Park HS. A practical data recovery technique for long-term strain monitoring of mega columns during construction. Sensors. 2013;13(8):10931–43.
27. Myers J, Bloch K. Comparison of prestress losses for pedestrian bridges constructed with high-strength concrete and high-strength self-consolidating concrete. J Bridge Eng. 2013;18(9):871–8.
28. Liu Y, Deng Y, Cai CS. Deflection monitoring and assessment for a suspension bridge using a connected pipe system: a case study in China. Struct Control Health Monit. 2015;22(12):1408–25.

29. Ministry of Transport of the People's Republic of China. Specification for design highway suspension bridge JTG TD65-05-2015. Beijing: China Communications Press; 2015.
30. Haight R, Billington D, Khazem D. Cable safety factors for four suspension bridge. J Bridge Eng. 1997;2(4):157–67.
31. Ministry of Transport of the People's Republic of China. General code for design highway bridges and culverts JTG D60-2004. Beijing: China Communications Press; 2004.
32. Gimsing NJ, Georgakis CT. Cable supported bridges, concept & design. 3rd ed. New York: Wiley; 2012.

Chapter 10
WIM-Based Fatigue Performance Investigation of Hangers

10.1 Introduction

For suspension bridges, hangers/suspenders are key load-carrying members, which connect main cables with stiffening girders. Dead loads from stiffening girders and the deck together with the transient live load are transferred to the main cables through hangers. The cyclic fluctuations in tension stresses induced by vehicle loads and wind loads can cause fatigue damage to the hangers [1, 2]. Therefore, considerable efforts have been devoted to the fatigue strength and life assessment of steel cables or hangers. For example, Takena et al. presented fatigue tests on a parallel-wire large-diameter cable system for cable-stayed bridges and a detailed review of other fatigue test results [3]. Suh and Chang used the Latin Square Design Method in the fatigue tests and evaluate the axial fatigue behavior of wire ropes used as hangers in suspension bridges [4]. Li et al. presented an empirical modeling of the deterioration process of steel wires in cables with consideration of the simultaneous occurrence of corrosion and fatigue [5]. However, very few studies have been reported on the fatigue performance investigation of hangers of in-service long-span suspension bridges, which is essential to decision making of inspection and replacement in bridge maintenance.

In order to properly carry out fatigue investigation for hangers, a reliable and efficient method to obtain the cyclic tension stress time histories is first needed. Recently, advanced sensing techniques have been developed to directly measure tension forces of cables or hangers. For example, He et al. presented a study on cable force monitoring by combining optical FBG sensors and BOTDA/R sensing technique in one single optical fiber [6]. Bao et al. proposed a new approach to identify the time-varying cable tension forces based on an adaptive sparse time–frequency analysis method [7]. Feng et al. proposed an accurate and cost-effective alternative for the determination of cable tension forces based on novel noncontact vision-based sensor technique [8]. These techniques had demonstrated their capacity in monitoring the varying tension forces of bridge cables or hangers. However, direct applications

Y. Deng and A. Li, *Structural Health Monitoring for Suspension Bridges*,
https://doi.org/10.1007/978-981-13-3347-7_10

of these sensing techniques for fatigue investigation of hangers have been rarely reported.

As an alternative, it presents a feasible option by taking full advantage of the information collected by structural health monitoring (SHM) systems for fatigue investigation of bridge hangers. Recently, traffic loading data collected by the weigh-in-motion (WIM) system, part of the bridge SHM system, has been used to conduct fatigue analysis for welded details in steel bridges through finite element (FE) simulation [9, 10]. The same method that integrates WIM measurements and FE simulation began to be used in fatigue investigation for bridge cables or hangers [2, 11, 12]. For example, Liu et al. investigated the influence of using rigid central clamps on the fatigue performance of short suspenders. They proposed a fatigue life evaluation procedure based on a vehicle load model and the wind spectrum developed from field-monitored data [12]. In the above studies, the probabilistic distributions of the parameters of vehicle loads were established by using the WIM measurements. Then, the random vehicle flows, generated using the Monte Carlo method, were applied to the FE models to estimate the time histories of the cyclic stress of the cables, hangers or welded details. The advantage of this method is that relatively small computation consumption can effectively cover the effects of the vehicle loads on the fatigue performance of the bridge components. However, this method cannot consider the time-varying nature of actual vehicular loads. Furthermore, the time-dependent features of the structural responses also get ignored in the aforementioned calculation processes.

Petrini and Bontempi estimated the fatigue life for long-span suspension bridge hangers under wind action and train transit. The results revealed that the main contribution of the fatigue damage of hangers is from train transit [13]. In fact, most of the existing studies indicated that the traffic load is the main cause of the fatigue damage of bridge suspenders or hangers [1, 2, 13].

In this chapter, a framework of fatigue performance investigation for hangers of long-span suspension bridges is proposed. The framework is developed by integrating the WIM measurements with the FE simulation. Differing from most existing methods, in this study, the full information of the vehicle loads provided by the WIM system is used to reconstruct real vehicular loading flows. The tension force time histories of hangers are calculated day-by-day by applying the reconstructed vehicle loadings to the influence surfaces that are extracted from FE analysis. Eight-month WIM measurements for a prototype suspension bridge are processed and analyzed. Critical calculation details are discussed. The time-dependent features of fatigue damages for hangers are investigated, and subsequently the fatigue lives are predicted. Finally, the correlation between the fatigue damage and the WIM data is investigated.

10.2 Proposed Framework of Fatigue Performance Investigation of Hangers

The fatigue performance investigation framework for the hangers of suspension bridges is proposed first in this chapter. The framework is illustrated in Fig. 10.1. The full information of the bridge-site WIM measurements is involved in this framework and plays an important role in the fatigue performance investigation of the hangers. The technical details of each step will be addressed in the following contents. The outlines of the framework are given as follows:

Step 1. Development of influence surfaces for hangers. A refined FE model including 3D modeling of the suspension bridge deck should be established. Then the influence surface rather than the influence line is calculated for hangers' tension forces. Linear interpolation is recommended to obtain continuous influence surfaces from the discrete FE analysis results.

Step 2. Reconstruction of vehicular loading flows based on WIM measurements. The time interval of the vehicle loading flow reconstruction is set to be one day. Based on the information of arriving time and speed, the location of each vehicle at 00:00:00

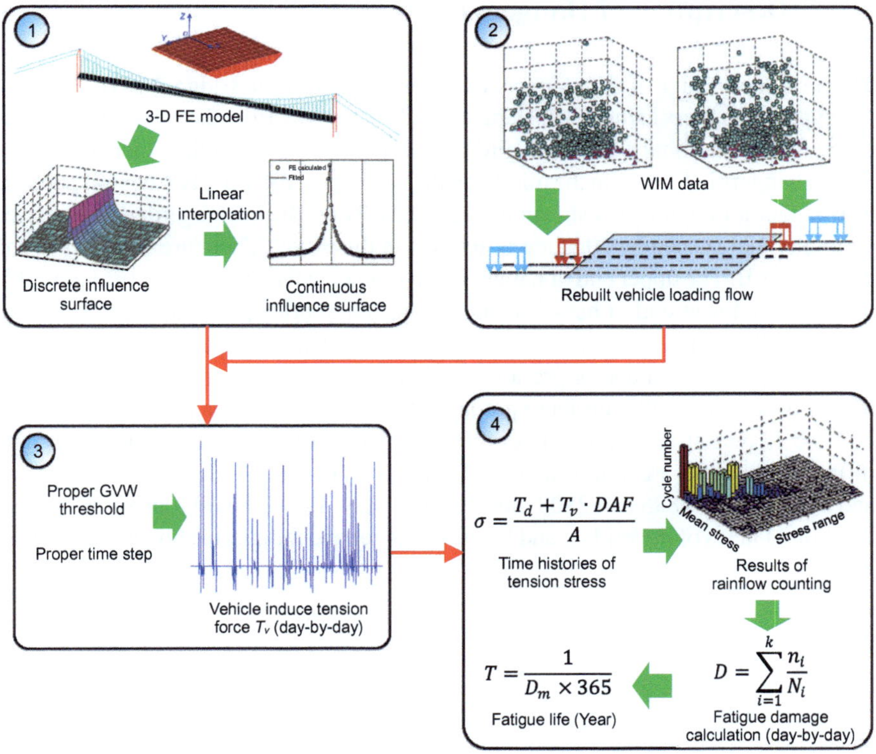

Fig. 10.1 Conceptual framework for fatigue performance investigation

can be determined. The speed and lane location are assumed to be constant from 00:00:00 until the vehicle leaves the bridge. Each n-axle vehicle can be simplified into n concentrated forces at their axle locations. The loading positions of the concentrated forces are on the centerline of the lane.

Step 3. Calculation of time histories of hanger tension forces. For saving computational effort, a gross vehicle weight (GVW) threshold is set to eliminate the lightweight vehicles that contribute little to the fatigue damage of hangers. Then, the vehicle loading flows of all lanes are applied to the influence surfaces obtained in step 1. It is noted that a proper time step should be chosen to ensure both computational efficiency and accuracy.

Step 4. Fatigue damage evaluation and life prediction. The vehicle dynamic effect is taken into account by multiplying the calculated hanger tension forces by a code-specified dynamic amplification factor. The rainflow counting algorithm is used to extract the stress cycles as well as the mean stress. Subsequently, the fatigue damage of the hangers can be calculated based on the Miner ruler with consideration of the effect of the mean stress. Finally, the fatigue lives of all hangers can be predicted based on the fatigue damage evaluation results.

10.3 Description of Hangers of the NSB

The prototype bridge of this chapter is the Nanxi Suspension Bridge (NSB). There are two types of hangers in the NSB. Hanger A is 13.2 m away from the center of the tower along the longitudinal axis of the girder. The distance between the other two neighboring hangers is 12.8 m. In the middle of the main span, there are two central buckles. The hanger arrangement is shown in Fig. 10.2a. Thus, the bridge girder is supported by totally 124 suspenders that are of interest in this study. There are 4 hanger A and 120 hanger B. The upper end of hanger is pin-connected to main cables through cable clamps. The lower end of hanger is pin-connected to bridge deck as well. The cross sections of hanger A and hang B are shown in Figs. 10.2b, c, respectively. Both hanger A and hanger B are made of prefabricated parallel wire strands (PPWS) and 8 mm thick Polyethylene (PE) protecting coat. Hanger A and hanger B are composed of 127 and 73 wire strands, respectively. The diameter and the nominal tension strength of the wire strand are 7 mm and 1670 MPa, respectively. The areas of hanger A and hanger B are 4885 and 2808 mm^2, respectively. The standard transverse section, location of lanes of the bridge, and location hanger anchorage are shown in Fig. 10.3.

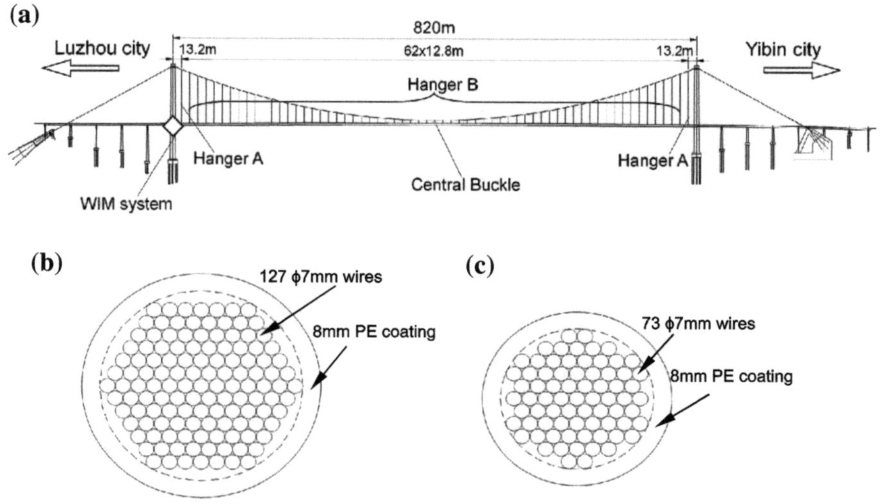

Fig. 10.2 Hangers of the NSB: **a** arrangement of the hangers, **b** cross-section of hanger A and **c** cross section of hanger B

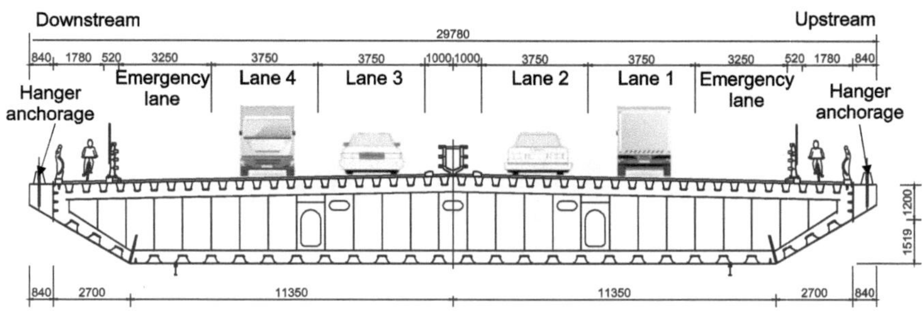

Fig. 10.3 Cross section of the main girder (Unit: mm)

10.4 WIM Measurements of the NSB

The WIM sensors were installed in the pavement at the location of the Luzhou tower, as shown in Fig. 7.3. There are six lanes on the deck while the outermost two lanes are for emergency. Hence, lane 1, lane 2, lane 3, and lane 4 were mounted with WIM sensors to monitor the in-service information of vehicle load. The WIM sensors can collect the vehicle loading data, such as arriving time, vehicle type, lane location, speed, GVW, vehicle length, and axle weights. As shown in Fig. 7.3, the vehicle driving direction of lane 1 and lane 2 is from Yibin to Luzhou. The driving direction of lane 3 and lane 4 is the opposite. In China, lane 1 and lane 4 are designed as slow

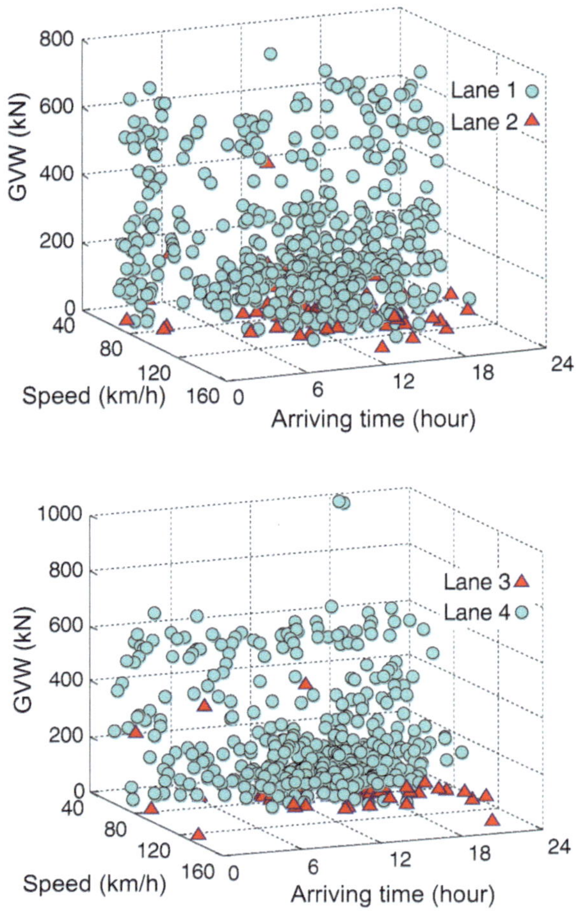

Fig. 10.4 Actual vehicle flows on January 01, 2014

lanes, which are mainly for trucks with lower speed. Lane 2 and lane 3 are fast lanes for high-speed vehicles.

Figure 10.4 visualizes the vehicle flow measured by WIM sensors on January 01, 2014, with vehicle information including lane location, speed, arriving time, and GVW. The arriving time refers to the time that a vehicle arrives at the WIM sensors. The vehicles with weight lower than 30 kN are eliminated from the vehicle flows. The rationale of eliminating the lightweight vehicle will be discussed later. There are a total of 1073 trucks passing through the WIM sensors on this selected day. The total traffic flow can be divided into 4 groups in terms of the lane location, i.e., Lane 1, lane 2, lane 3, and lane 4 with, respectively, 488, 78, 88, and 419 trucks. The figure shows that the weight of most vehicles on the fast lanes (lane 2 and lane 3) is below 100 kN, while the trucks in the slow lanes (lane 1 and lane 4) are heavier than 100 kN. In addition, the averaged driving speeds of lanes 1 through 4 are 76.5, 95.0, 91.0, and 74.4 km/h, respectively. Therefore, it can be concluded from this one-day WIM data that lane 1 and lane 4 have heavier vehicle loads and lower driving speed than those of lane 2 and lane 3.

To further investigate the variations of the vehicle loads on the NSB, long-term WIM data measured from January to August in 2014 are analyzed. Let f, f_1, f_2, f_3, and f_4 denote, respectively, the total vehicle number, vehicle numbers of lane 1, lane 2, lane 3, and lane 4. Then the equivalent GVW can be defined as

$$w = \left(\frac{1}{f}\sum_{l=1}^{f} GVW_l^m\right)^{\frac{1}{m}} \tag{10.1}$$

$$w_i = \left(\frac{1}{f_i}\sum_{k=1}^{f_i} GVW_{k,i}^m\right)^{\frac{1}{m}} \quad (i = 1, 2, 3, 4) \tag{10.2}$$

where w is the equivalent GVW of all lanes, w_i the equivalent GVW of lane i, GVW_l the GVW of the lth vehicle, $GVW_{k,\,i}$ the GVW of the kth vehicle of lane i, and m the slope of the fatigue S-N curve of the hangers. Herein, the value of m is set to 3.5 (see Eq. (10.7)). In addition, the vehicle loading factors are defined as

$$\alpha = f w^m \tag{10.3}$$

$$\alpha_i = f_i w_i^m (i = 1, 2, 3, 4) \tag{10.4}$$

where α and α_i are the vehicle loading factors of all lanes and lane i, respectively.

Figure 10.5 compares the defined factors that represent vehicle loads on the NSB. One month is taken as the time interval to calculate these factors. Figure 10.5a presents the daily averaged values of the vehicle number. In general, all factors increase gradually in the first eight months of 2014. As shown in Figs. 10.5b, c, the equivalent GVW and vehicle loading factor of the slow lanes (lanes 1 and 4) are much larger than those of the fast lanes (lanes 2 and 3), mainly due to the reason that more truck vehicles are running on the slow lanes.

Furthermore, differences between factors of lane 1 (slow lane in the upstream) and lane 4 (slow lane in the downstream) can be observed. The vehicle number, equivalent GVW, and vehicle loading factor of lane 4 were slightly smaller than those of lane 1 from January to June. Then the relationship was reversed in July and August. The factors defined in this chapter will be used later for correlation analysis.

Fig. 10.5 Vehicle load from January to August of 2014: **a** vehicle number, **b** equivalent GVW and **c** vehicle loading factor

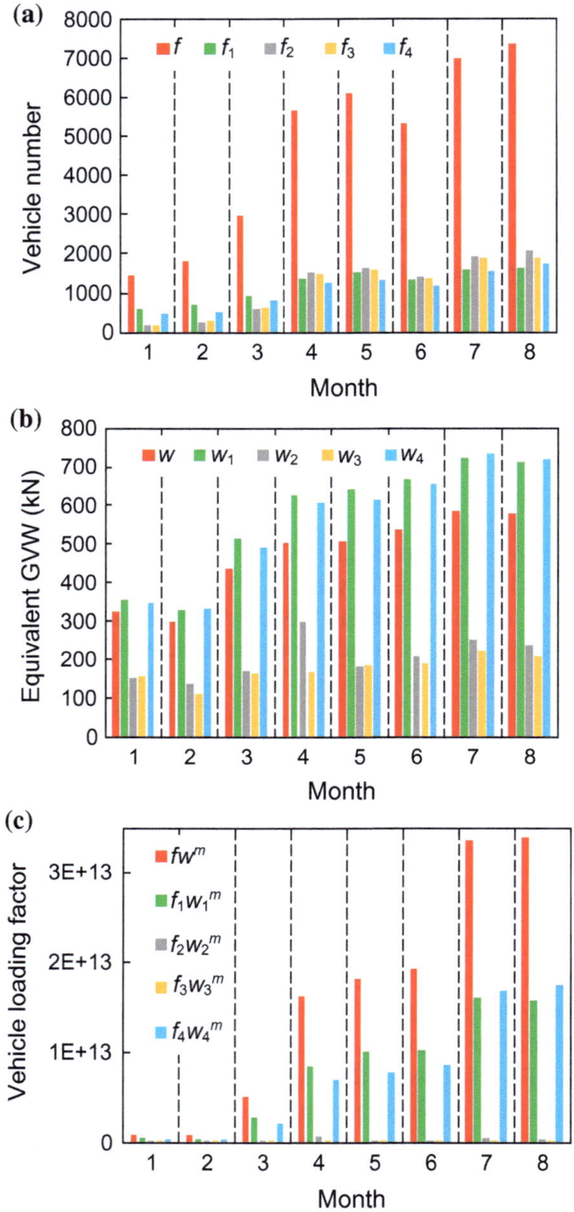

10.5 Fatigue Performance Investigation of the NSB's Hangers

10.5.1 Influence Surfaces of Hanger Tension Force

According to the above WIM data analysis, the lane vehicle load differs from one other. The influence lines for hangers' tension forces cannot consider the transverse distribution of vehicular loads. Hence, the influence surface method is used to account for the effects of lane loading variations. Refined three-dimensional (3D) FE model is needed for calculating influence surfaces. In Chap. 7, a simple FE model of the NSB using the commercial ANSYS software was established and used to verify a deflection monitoring system, in which the main girder was simulated using beam elements. FE model updating and validation were performed based on dynamic and deflection testing results from a controlled loading test. FE analysis results were in excellent agreement with the dynamic properties and deflection measurement results. The details of the modeling, updating, and validation of the FE model can be found in the literature [14]. Based on the above beam-girder FE model, a more refined FE model is developed by using shell elements to model the main girder of the NSB, as shown in Fig. 10.6. Shell 63 element is adopted for the orthotropic steel box girder, Link 8 for the main cables and hangers, and Beam 4 for the central buckles and the bridge towers. As shown in Fig. 10.6, in the bridge downstream, the hanger is numbered from H1 to H62 (from Luzhou to Yibin), and in the bridge upstream, the

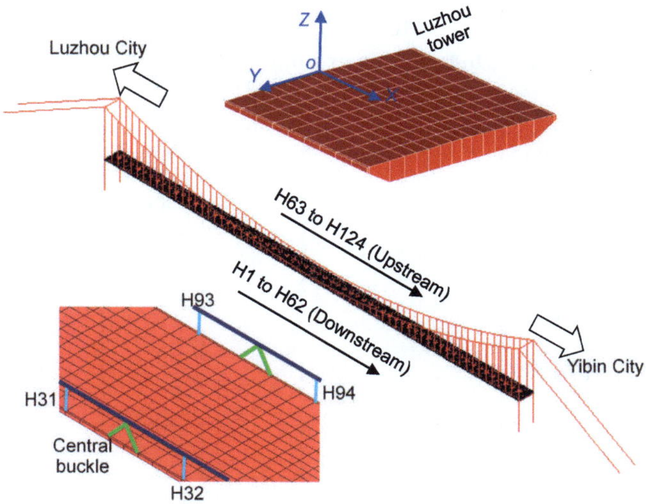

Fig. 10.6 Finite element model of the NSB

hanger is numbered from H63 to H124 (from Luzhou to Yibin). The controlled load test results of the NSB including dynamic properties and vertical deflection of the main girder can be found in Chap. 7. The controlled load test was carried out after bridge completion. The FE model of the NSB in Fig. 10.6 has also been validated by using the results of the controlled load test. The shell element-based model is in better agreement with the results of the controlled load test than the beam element-based model.

Since the purpose of the refined FE model is to determine the practical transverse locations of the passing vehicles rather than analyzing structural behaviors for the steel box girder, it is not necessary to fully model all members of the orthotropic steel box girder. As shown in Fig. 10.6, although U ribs are not modeled, the stiffness strengthening of U ribs to the deck and bottom plates are taken into account. The deck and bottom plates of the box girder are modeled using the orthotropic shell element, of which the Young's Modulus E_X and E_Y are different. The FE model in Fig. 10.6

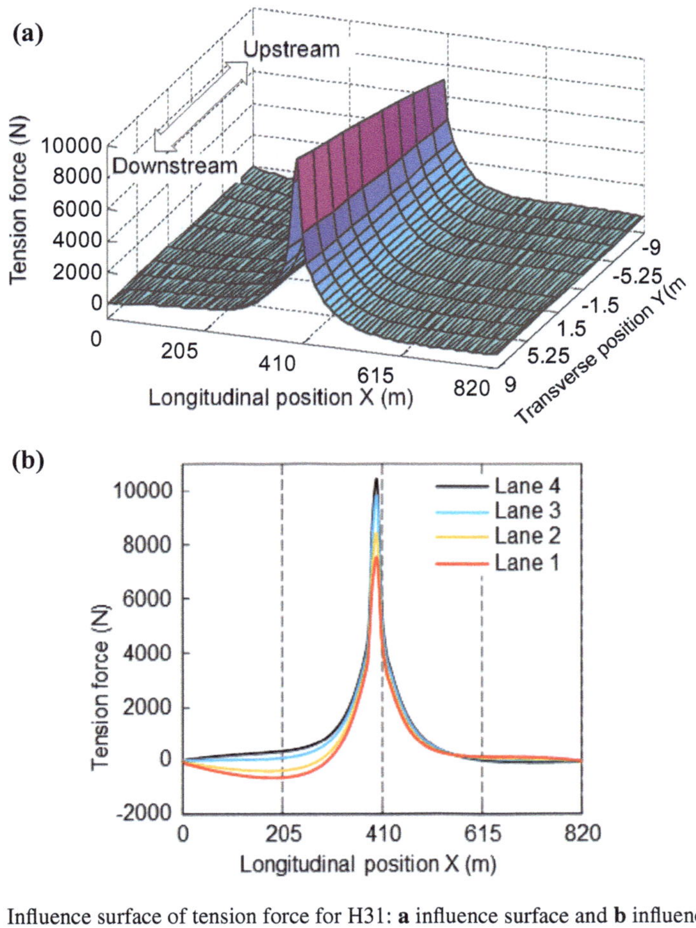

Fig. 10.7 Influence surface of tension force for H31: **a** influence surface and **b** influence lines

has a total of 9012 nodes and 13,311 elements. Considering the balance between calculation accuracy and model complexity, the mesh size of the shell elements for the main girder needs to be selected carefully. Along the longitudinal axis of the girder, the maximum length of the shell element is 1.6 m.

By applying a concentrated force 100 kN node-by-node on the deck plate, the influence surface of the tension force for each hanger can be derived. Figure 10.7a shows the influence surface for hanger H31. The peak of the tension force appears at the longitudinal position corresponding to hanger H31. Figure 10.7b further gives the influence lines of H31, with the transverse position of the influence line at the centerline of each lane. It is found that the maximum H31 tension forces for 100 kN loadings on Lane 1 through Lane 4 are 7.5, 8.4, 9.7, and 10.4 kN, respectively.

The influence surface in Fig. 10.7 are plotted by connecting discrete points of tension forces obtained from FE calculation. In order to calculate the tension force under the real vehicle load, the continuous influence surfaces are obtained by using linear

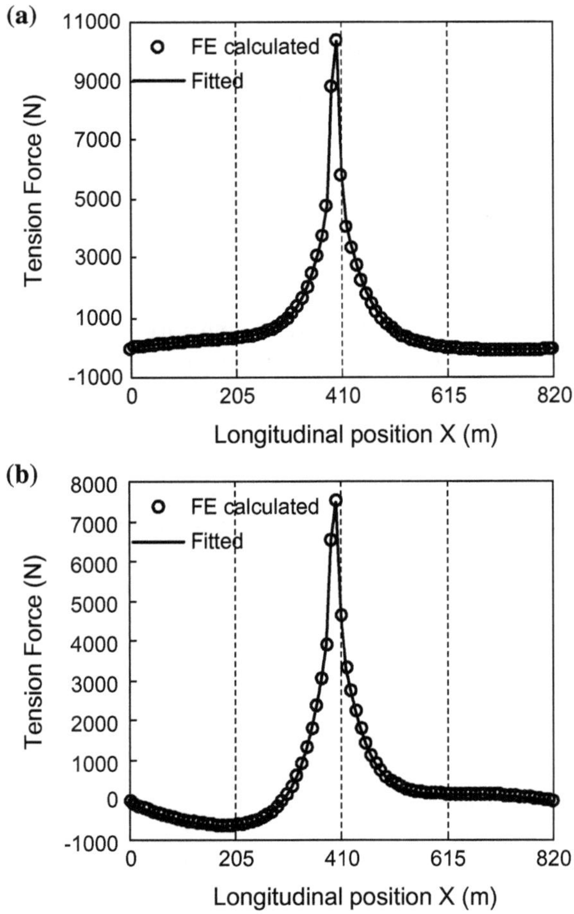

Fig. 10.8 Comparison between FE-calculated and fitted influence surface of H31: **a** centerline of lane 4 and **b** centerline of lane 1

interpolation to fit the discrete influence surface as plotted in Fig. 10.7. Figure 10.8 shows the FE-calculated and fitted influence lines of lane 1 and lane 4. For centerline of lane 4, the FE-calculated and fitted maximum tension forces are 10391 and 10326 N, respectively. For centerline of lane 1, the FE-calculated and fitted maximum tension forces are 7512 and 7471 N, respectively. The results in Fig. 10.8 show that the calculated and fitted influence surfaces are in great agreement.

10.5.2 Vehicle Loading Configuration

As presented in the previous chapter, the practical vehicle loading flows can be reconstructed based on the full information of each passing vehicle recorded by WIM sensors. First, one day is chosen as the time interval for WIM data processing. The speed and lane location of each vehicle is assumed to be constants. Second, the location of each vehicle at 00:00:00 of each day can be determined by using arrival time and speed. Thus, four vehicle sequences representing the four-lane traffic flows are developed. Third, the vehicle sequences are applied to the continuous influence surface of each hanger in MATLAB. In this manner, the time history of tension force for each hanger from 00:00:00 to 24:00:00 can be obtained.

In the above implementation, the loading configuration of each vehicle should be considered. Taking July 23, 2014 as an example, a four-axle truck which arrived at the WIM system at 00:02:13. The lane location, speed and GVW of this truck are lane 1, 65 km/h, and 717 kN, respectively. Figure 10.9 illustrates the passing process of this truck. The time of this truck arriving at Yibin tower and hanger H31 were 00:01:27 and 00:01:51 with the assumption of a constant moving speed. Noting that during this period there was no other truck passing the bridge. With the information of vehicle type and axle weights, the original loading configuration of this truck is shown in Fig. 10.10. Eight concentrated forces represent the weight of the eight-wheel patches. In addition, Fig. 10.10 shows the transverse location of the truck. When the truck passing through the bridge deck, the distance between the truck centerline and the centerline of lane 1 is assumed to vary in a 1.2 m range. For calculation simplicity, the simplified loading configuration, including four concentrated forces,

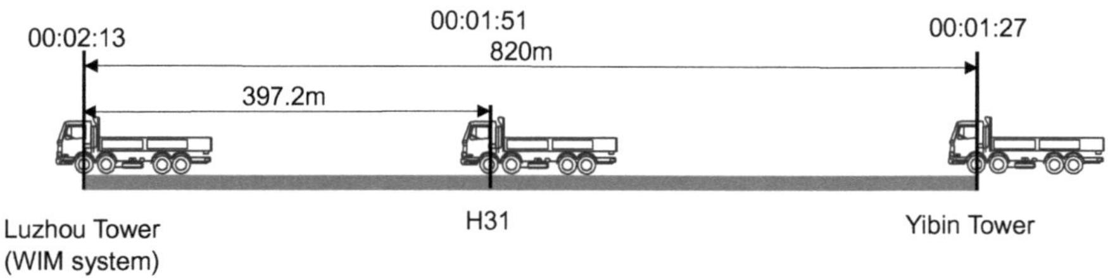

Fig. 10.9 A 4-axle truck passing through the bridge deck

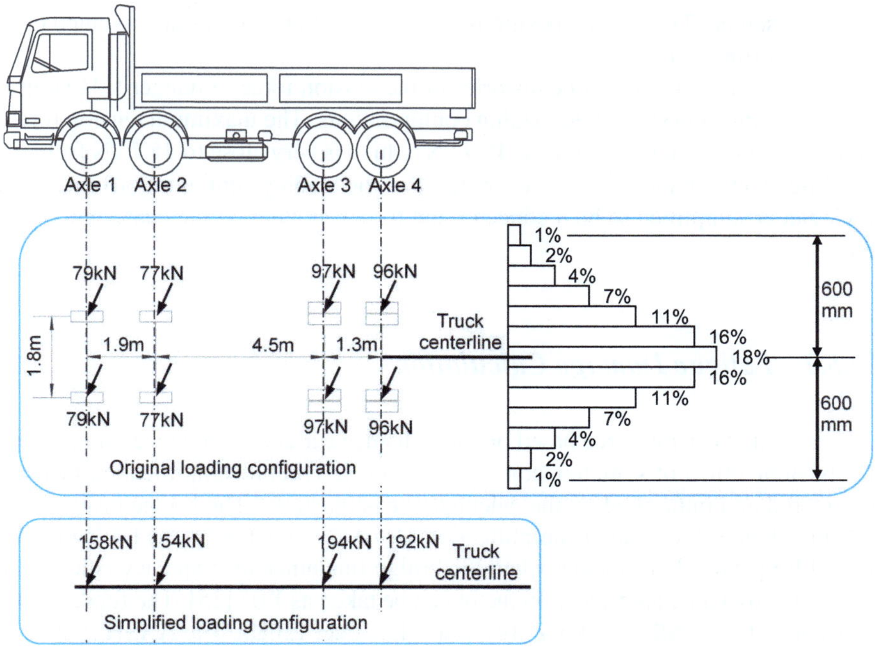

Fig. 10.10 Loading configuration of a four-axle truck

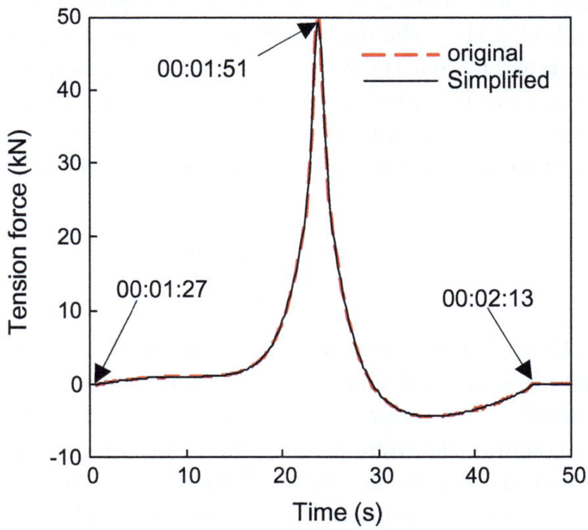

Fig. 10.11 Tension force time history induce by a 4-axle truck

is also presented. The loading positions of the concentrated forces are assumed to be at the centerline of lane 1.

Figure 10.11 compares time histories of the tension force of hanger H31 subject to both original and simplified loading configurations. The maximum tension forces from the original configuration are 45.9 kN, which is very close to 45.7 kN from the simplified configuration. The results reveal that the loading configuration of a n-axle truck can be simplified to be n concentrated forces moving on the centerline of the lane.

10.5.3 Fatigue Damage Calculation

The calculated tension force based on the influence surface cannot take into account the dynamic effect of vehicle loads. In this chapter, dynamic amplification factor is considered and multiplied by the calculated tension stress. The bridge design code in China defines the dynamic amplification factor as a function of the bridge fundamental frequency. It states that when the bridge fundamental frequency is less than 1.5 Hz, the dynamic amplification factor can be taken as 1.05 [15]. The fundamental frequency of the NSB is 0.18 Hz [14], much less than 1.5 Hz. The AASHTO LRFD proposed that the dynamic load allowance (impact factor) for all components except deck joints under fatigue limit state is 0.15 [16]. Ma et al. adopted the vehicle-bridge coupling vibration analysis to investigate the dynamic amplification factor of a long-span suspension bridge [17]. The main span and the fundamental frequency of the bridge are 1386 m and 0.0868 Hz, respectively. It was found that their results are closer to the provision of the AASHTO LRFD than the Chinese bridge design code. On this basis, the dynamic amplification factor for the fatigue analysis is conservatively set to 1.15 in this chapter. Thus, the tension stress σ of the hanger can be defined as

$$\sigma = \frac{T_d + T_v \cdot DAF}{A} \tag{10.5}$$

where T_d is the hanger tension force induced by deal load, T_v, is the tension force induced by vehicle loads and calculated by using the influence surface method, DAF is the dynamic amplification factor. A is the cross-sectional area of the hanger.

The rainflow counting algorithm [18] is utilized to process the time history of the tension stress σ. Then with the extracted stress cycles, the fatigue damage of suspension bridge hangers is estimated by using the Miner ruler [19]

$$D = \sum_{i=1}^{k} \frac{n_i}{N_i} \tag{10.6}$$

where n_i is the cycle number of the stress range S_i, N_i the fatigue life when the stress range is S_i, D the fatigue damage. The S-N curves of the hangers are obtained from the previous studies [20, 21]

$$\log N = 13.95 - 3.5\log S \tag{10.7}$$

Previous studies suggested that the fatigue life of cables is a function of mean stress level as well as the stress range [4]. In this chapter, the effects of mean stress on the fatigue damage are considered by using Goodman diagram [22]. The modified stress range S_m is

$$S_m = \frac{S}{1 - \frac{\sigma_m}{\sigma_u}} \tag{10.8}$$

where σ_m and σ_u are the mean stress and ultimate tension strength, respectively. Thus, the stress range S in Eq. (10.7) should be replaced by S_m.

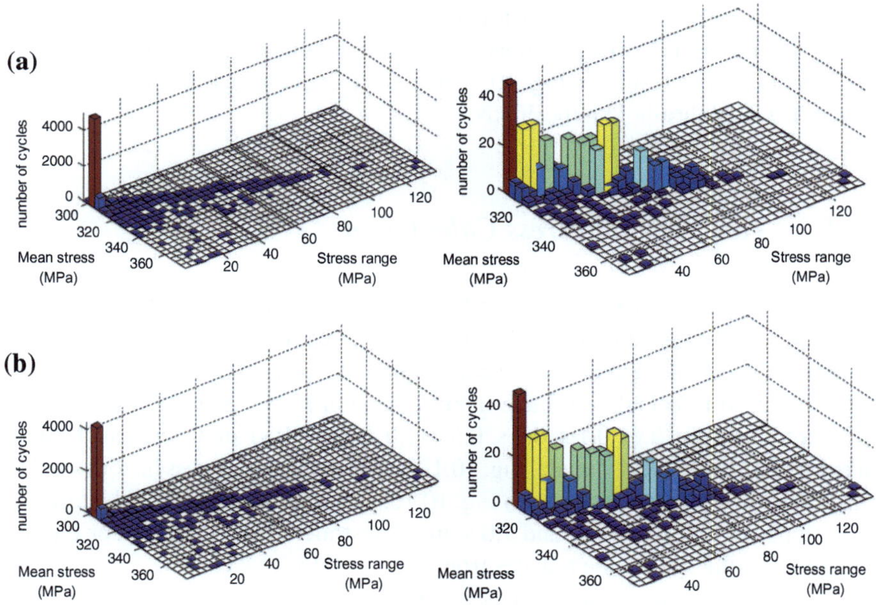

Fig. 10.12 Rainflow counting histograms of H31 on July 23, 2014: **a** without threshold and **b** with threshold

Table 10.1 Fatigue damage on July 23, 2014

Hanger no.	Without threshold	With threshold	Relative error (%)
H1	7.2396E−07	7.2363E−07	0.05
H15	1.2711E−05	1.2703E−05	0.07
H31	1.8350E−05	1.8318E−05	0.17

Note Fatigue damage relative error is given by: relative error (%) $= \left|\frac{D_{\text{without}}-D_{\text{with}}}{D_{\text{without}}}\right| \times 100\%$, where D_{without} and D_{with} mean the fatigue damage without threshold and with threshold, respectively

10.5.4 Threshold of the GVW

The WIM data analysis reveals that the vehicle number increased significantly from January to August in 2014. However, the GVWs of a large proportion of the vehicles are less than 30kN. Taking WIM data of July 23, 2014 as an example, there are a total of 8758 vehicles while the GVW of 1412 vehicles is less than 30 kN. For computational efficiency, the GVW threshold is set to be 30 kN. Fig. 10.12 shows the rainflow counting histograms for hanger H31 on July 23, 2014, which indicate that the difference between the histograms with and without threshold is negligibly small. Moreover, Table 10.1 compares the fatigue damage results for hangers H1, H15 and H31. The results in Fig. 10.12 and Table 10.1 suggest that it is reasonable to set the GVW threshold to be 30 kN.

10.5.5 Time Step for Stress Calculation

The time step is a key issue in tension force calculation. If the time step is too large, the peak values of tension forces may not be captured. On the other hand, too short time step may result in unnecessary computational effort. In the above analysis, the time step used to calculate vehicle-induced tension forces is 0.2 s. The effects of different time steps (i.e., 0.1 s, 0.2 s, 0.5 s, 1.0 s, and 2.0 s) are investigated. Again, take July 23, 2014 as an example, Fig. 10.13 shows the tension stress histories of H31 subject to the 4-axle truck shown in Fig. 10.10. The peak values of time histories for time steps 0.1 s, 0.2 s, 0.5 s, and 1.0 s are almost identical. However, for the time step of 2.0 s, the peak value reduces from 325.5 to 313.2 MPa.

The rainflow counting algorithm is used to process the time histories in Fig. 10.13. The results are listed in Table 10.2. Two half-cycle can be extracted from each time history. Except for time step of 2.0 s, stress range and mean stress are very close for all other time steps. The results in Fig. 10.13 and Table 10.2 are based on a relatively small time-scale (about 15 s). When the time-scale is extended to one day, Fig. 10.14 presents the fatigue damages for hanger H15 and H31 on July 23, 2014 with regarding various time step. When the time step is 0.1 s, the fatigue damage of H15 and H31 converge to 1.28E−05 and 1.85E−05, which can be taken as the true values. When the time step is doubled to be 0.2 s, the fatigue damages of H15 and H31 slightly

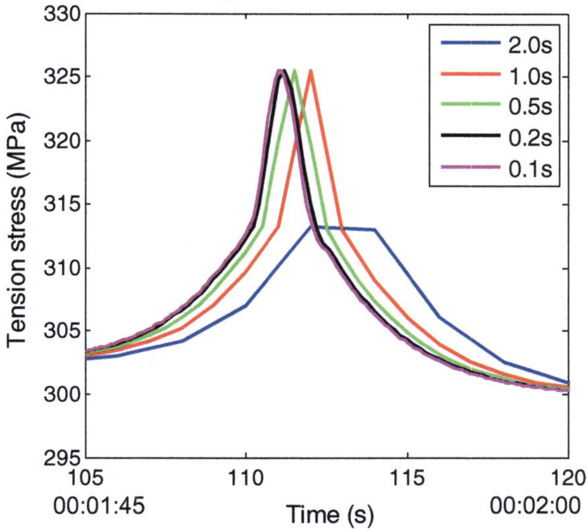

Fig. 10.13 Tension stress history of H31 on July 23, 2014

decrease to 1.27E−05 and 1.83E−05. Then, when the time step further increases, the fatigue damage errors become unacceptable. For example, when the time step is 2.0 s, the fatigue damages of hangers H15 and H31 are only 7.02E−06 and 8.84E−06. In other words, the fatigue damages for hangers H15 and H31 decrease by 45.3 and 52.2% when the time step increases from 0.1 to 2.0 s. To save computational effort, the time step is set to be 0.2 s.

10.5.6 Fatigue Life Prediction

The WIM data collected from January to August in 2014 is processed and used to compute the fatigue damage of all hangers of the NSB. Figure 10.15 presents the monthly averaged fatigue damage D_m for January and August.

$$D_m = \sum_{i=1}^{l} \frac{D_i}{l} \tag{10.9}$$

where l is the number of the days in the specific month; D_i the fatigue damage on the ith day. For example, l is 31 when it is January. At both the bridge downstream and upstream, hangers H29 and H34, which are symmetrical with respect to (w.r.t.) the mid-span, have the maximum fatigue damages, while H1 and H62, also symmetrical w.r.t. the mid-span, have the minimum fatigue damages. In addition, a zigzag distribution in the hangers' fatigue damage can be observed along the bridge longitudinal direction.

Table 10.2 Stress cycles of H31 00:01:45 to 00:02:00 on July 23, 2014

Stress cycles	Time step									
	2.0 s		1.0 s		0.5 s		0.2 s		0.1 s	
Stress range (MPa)	10.7	12.3	22.4	24.9	22.2	25.1	22.1	25.1	22.0	25.2
Mean stress (MPa)	307.9	307.1	314.3	313.0	314.4	312.9	314.4	312.9	314.5	312.9
Number of cycles	0.5	0.5	0.5	0.5	0.5	0.5	0.5	0.5	0.5	0.5

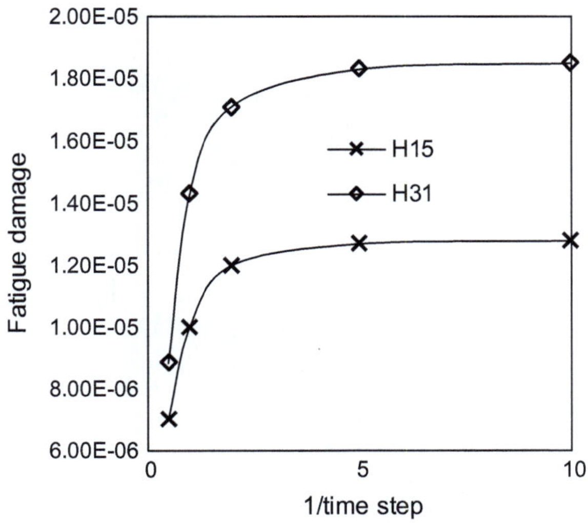

Fig. 10.14 Fatigue damage of H15 and H31 on July 23, 2014

A comparison shows that the fatigue damages increased almost two orders of magnitude from January to August. Four hangers are selected for further discussion. Hangers H31 and H93 are symmetrical w.r.t. the longitudinal bridge centerline. As shown in Fig. 10.16, the fatigue damage of H31 (downstream) is slightly smaller than that of H93 (upstream) from January to June, and slightly larger in July and August. The relationship between the downstream and upstream hangers' fatigue damages is found to be highly consistent with the relationship between the slow lanes' factors, as discussed in Sect. 10.4. Figure 10.16 also shows that H29 and H34, which are symmetrical w.r.t. the mid-span, have almost the same the fatigue damage from January to August in 2014.

Figures 10.17 and 10.18 presented the fatigue lives based on the monthly averaged fatigue damage in January and August, respectively. The fatigue life T of the hanger can be estimated by:

$$T = \frac{1}{D_m \times 365} \tag{10.10}$$

where D_m, which is the monthly averaged fatigue damage of a hanger, is shown in Fig. 10.15.

A decreasing trend is clearly observed for fatigue lives of hangers from both bridge towers to the bridge mid-span. When the monthly averaged fatigue damage in January is input into Eq. (10.10), fatigue lives of hangers are incredibly long. For example, the fatigue lives of hangers H29 and H31 are more than 5000 years. However, when the monthly averaged fatigue damage in August is used, hanger fatigue lives will be significantly shortened, which are less than 200 years. As discussed in Sect. 10.4, during the period from January to August in 2014, the vehicle number, equivalent GVW, and vehicle loading factor had greatly increased. It is also found that hangers

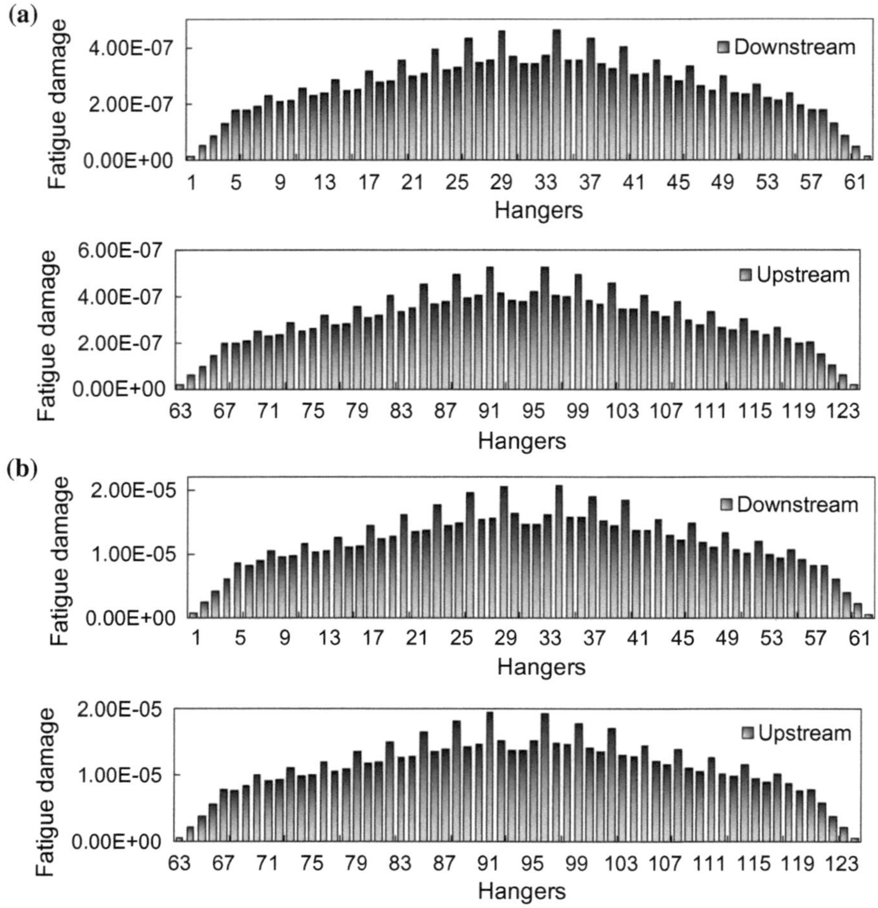

Fig. 10.15 Monthly averaged fatigue damage in **a** January and **b** August of 2014

with the shortest fatigue lives are not all those closest to the mid-span, i.e., H31, H31, H91, and H92. Instead, hangers H29, H34, H91, and H96 have the shortest fatigue lives.

10.5.7 Correlation Analysis Between the Fatigue Damage and Vehicular Loads

H29 has the largest fatigue damage over the selected operational period. Herein, H29 is selected to conduct correlation analysis between the hanger fatigue damage and vehicular loads. Figure 10.19a–c present the correlations as well as the linear fitting

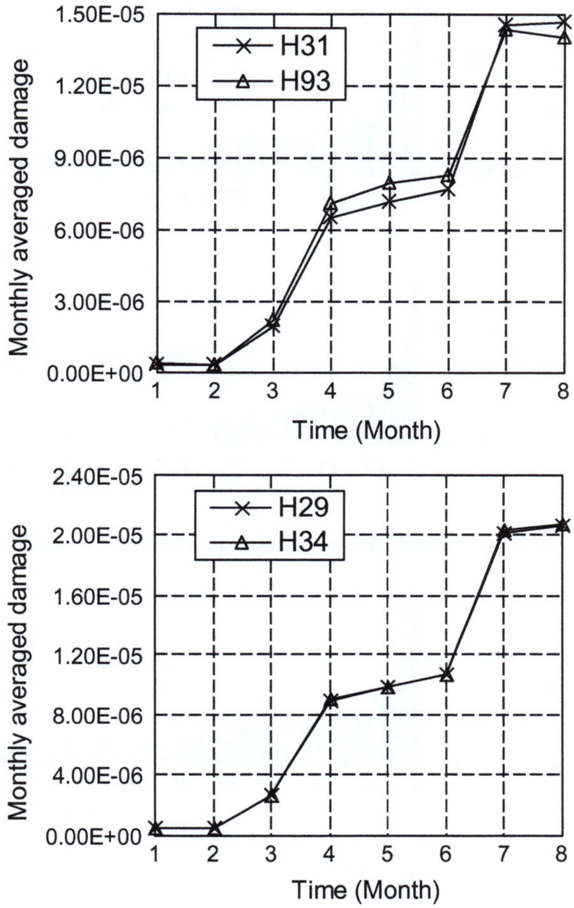

Fig. 10.16 Monthly averaged fatigue damage

formulas and correlation coefficients between the fatigue damage of hanger H29 and the vehicle number, the equivalent GVW, and the vehicle loading factor, respectively. The best correlation is found between the fatigue damage and the vehicle loading factor, then vehicle number, and lastly the GVW.

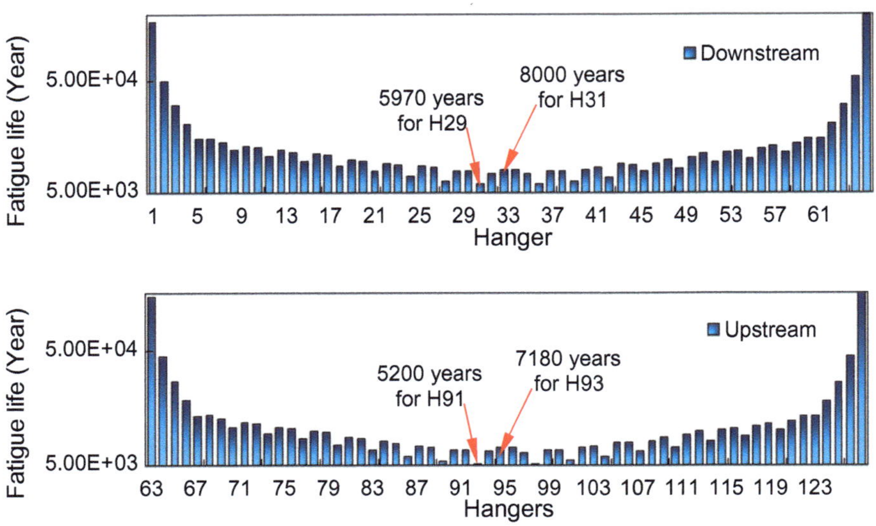

Fig. 10.17 Fatigue lives derived based on monthly averaged fatigue damage in January 2014

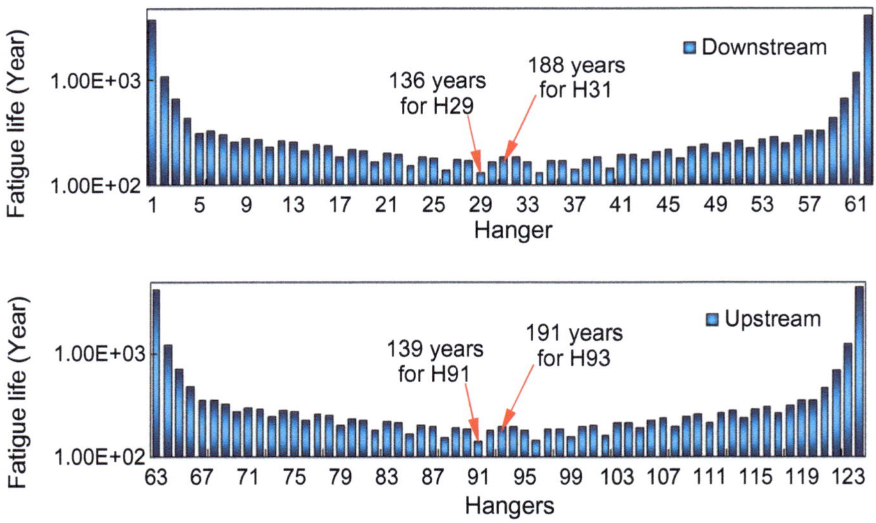

Fig. 10.18 Fatigue lives derived based on monthly averaged fatigue damage in August 2014

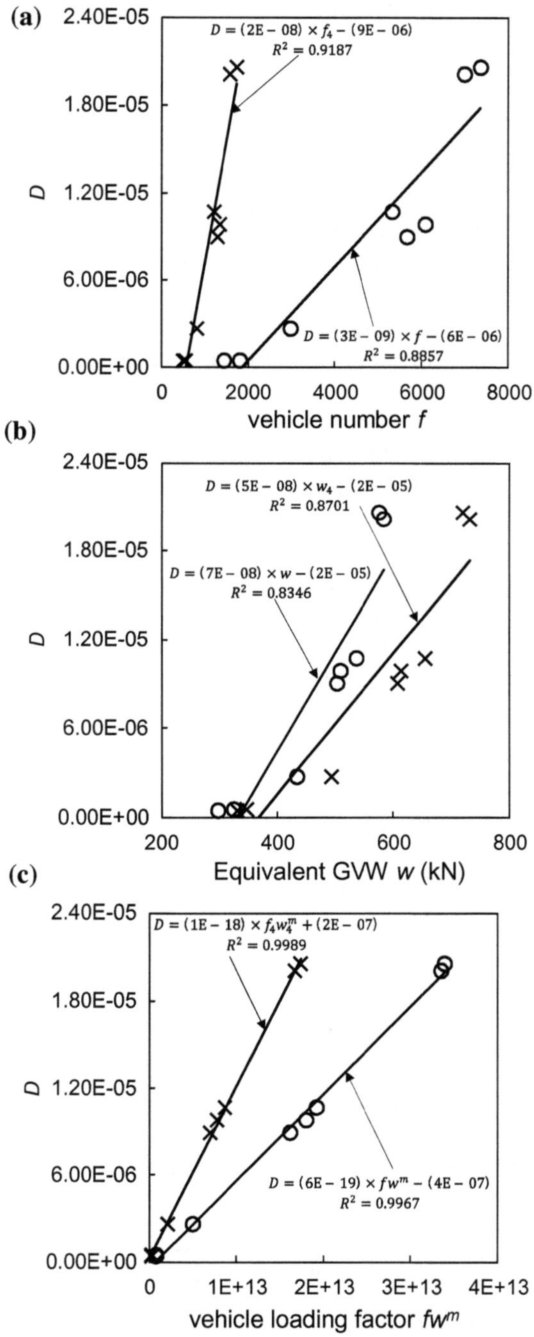

Fig. 10.19 Correlations between fatigue damage of H29 and vehicle loads: **a** vehicle number, **b** equivalent GVW and **c** vehicle loading factor

10.6 Summary

A framework is proposed to investigate the fatigue performance of hangers of long-span suspension bridges subject to time-varying and site-specific vehicle loadings. Following suggestions and conclusions can be drawn through studies for a prototype suspension bridge based on eight-month WIM data:

1. In order to consider the transverse distribution of actual vehicle loads, refined 3D FE model should be developed for the main girder of the suspension bridge. Linear interpolation is recommended to obtain the continuous functions from the discrete influence surfaces from FE analysis.
2. For computational efficiency, three measures are recommended: simplify each n-axle vehicle into n concentrated forces at their axle locations and applied on the centerline of the lane; set a GVW threshold (30 kN) to ignore loadings from lightweight vehicles since the differences between fatigue damages with and without threshold is pretty small; set a proper time step during stress calculation (herein, 0.2 s).
3. Fatigue lives of hangers have a decreasing trend from both bridge towers to the mid-span. Evaluated hanger fatigue damages and predicted fatigue lives significantly vary over time. Therefore, hanger fatigue performance should be studied based on long-term site-specific vehicle loadings.
4. Correlation analysis shows that the fatigue damage is linearly correlated to the vehicle loading factor, then vehicle number, and lastly the GVW.

References

1. Li SL, Zhu SY, Xu YL, Chen ZW, Li H. Long-term condition assessment of suspenders under traffic loads based on structural monitoring system: application to the Tsing Ma Bridge. Struct Control Health Monit. 2012;19(1):82–101.
2. Chen B, Li XZ, Xie X, Zhong Z, Lu PZ. Fatigue performance assessment of composite arch bridge suspenders based on actual vehicle loads. Shock Vib. 2015. 659092. https://doi.org/10.1155/2015/659092.
3. Takena K, Miki C, Shimokawa H, Sakamoto K. Fatigue resistance of large-diameter cable for cable stayed bridges. J Struct Eng-ASCE. 1992;118(3):701–15.
4. Suh JI, Chang SP. Experimental study on fatigue behaviour of wire ropes. Int J Fatigue. 2000;22(4):339–47.
5. Li SL, Xu Y, Zhu SY, Guan XC, Bao YQ. Probabilistic deterioration model of high-strength steel wires and its application to bridge cables. Struct Infrastruct Eng. 2015;11(9):1240–9.
6. He JP, Zhou Z, Ou JP. Optic fiber sensor-based smart bridge cable with functionality of self-sensing. Mech Syst Signal Process. 2013;35(1–2):84–94.
7. Bao YQ, Shi ZQ, Beck JL, Li H, Hou TY. Identification of time-varying cable tension forces based on adaptive sparse time-frequency analysis of cable vibrations. Struct Control Health Monit. 2017;24(3):e1889.
8. Feng DM, Scarangello T, Feng MQ, Ye Q. Cable tension force estimate using novel noncontact vision-based sensor. Measurement. 2017;99:44–52.

9. Guo T, Frangopol DM, Chen Y. Fatigue reliability assessment of steel bridge details integrating weigh-in-motion data and probabilistic finite element analysis. Comput Struct. 2012;112:245–57.
10. Liu Y, Zhang HP, Liu YM, Deng Y, Jiang N, Lu NW. Fatigue reliability assessment for orthotropic steel deck details under traffic flow and temperature loading. Eng Fail Anal. 2017;71:179–94.
11. Liu ZX, Guo T, Chai S. Probabilistic fatigue life prediction of bridge cables based on multi-scaling and mesoscopic fracture mechanics. Appl Sci Basel. 2016;6(4):UNSP99.
12. Liu ZX, Guo T, Huang LY, Pan ZH. Fatigue life evaluation on short suspenders of long-span suspension bridge with central clamps. J Br Eng. 2017;22(10):04017074.
13. Petrini F, Bontempi F. Estimation of fatigue life for long span suspension bridge hangers under wind action and train transit. Struct Infrastruct Eng. 2011;7(7–8):491–507.
14. Liu Y, Deng Y, Cai CS. Deflection monitoring and assessment for a suspension bridge using a connected pipe system: a case study in China. Struct Control Health Monit. 2015;22(12):1408–25.
15. Ministry of Transport of the People's Republic of China. General code for design highway bridges and culverts JTG D60-2015. Beijing: China Communications Press; 2015.
16. AASHTO. AASHTO LRFD bridge design specifications, 6th ed. 2012.
17. Ma L, Han WS, Ji BH, Shi GG, Liu JX. Study of vehicle-bridge coupling vibration under actual traffic flow. China J Highw Transp. 2012;25(6):80–7.
18. Nieslony A. Determination of fragments of multiaxial service loading strongly influencing the fatigue of machine components. Mech Syst Signal Process. 2009;23(8):2712–21.
19. Miner MA. Cumulative damage in fatigue. J Appl Mech-Trans ASME. 1945;12(3):A159–64.
20. Faber MH, Engelund S, Rackwitz R. Aspects of parallel wire cable reliability. Struct Saf. 2003;25(2):201–25.
21. Zeng Y, Chen AR, Tan HM. Fatigue assessment of hanger wires of suspension bridges in its operation life based on in-situ traffic flow. J Disaster Prev Mitig Eng. 2014;34(2):185–91.
22. Hosford WF. Mechanical behavior of materials. 2nd ed. Cambridge University Press; 2005.

MIX
Papier aus verantwortungsvollen Quellen
Paper from responsible sources
FSC® C105338

If you have any concerns about our products,
you can contact us on
ProductSafety@springernature.com

In case Publisher is established outside the EU,
the EU authorized representative is:
Springer Nature Customer Service Center GmbH
Europaplatz 3, 69115 Heidelberg, Germany

Printed by Libri Plureos GmbH
in Hamburg, Germany